Synthesis Lectures on Communications

This series of short books cover a wide array of topics, current issues, and advances in key areas of wireless, optical, and wired communications. The series also focuses on fundamentals and tutorial surveys to enhance an understanding of communication theory and applications for engineers.

Rongrong Zhang · Hao Liu

RFID Applications

Secure and Efficient Backscatter Networking

 Springer

Rongrong Zhang
Capital Normal University
Beijing, China

Hao Liu
Communication University of China
Beijing, China

ISSN 1932-1244　　　　　　　　ISSN 1932-1708　(electronic)
Synthesis Lectures on Communications
ISBN 978-3-031-93033-1　　　ISBN 978-3-031-93034-8　(eBook)
https://doi.org/10.1007/978-3-031-93034-8

© The Editor(s) (if applicable) and The Author(s), under exclusive license to Springer Nature Switzerland AG 2026

This work is subject to copyright. All rights are solely and exclusively licensed by the Publisher, whether the whole or part of the material is concerned, specifically the rights of translation, reprinting, reuse of illustrations, recitation, broadcasting, reproduction on microfilms or in any other physical way, and transmission or information storage and retrieval, electronic adaptation, computer software, or by similar or dissimilar methodology now known or hereafter developed.
The use of general descriptive names, registered names, trademarks, service marks, etc. in this publication does not imply, even in the absence of a specific statement, that such names are exempt from the relevant protective laws and regulations and therefore free for general use.
The publisher, the authors and the editors are safe to assume that the advice and information in this book are believed to be true and accurate at the date of publication. Neither the publisher nor the authors or the editors give a warranty, expressed or implied, with respect to the material contained herein or for any errors or omissions that may have been made. The publisher remains neutral with regard to jurisdictional claims in published maps and institutional affiliations.

This Springer imprint is published by the registered company Springer Nature Switzerland AG
The registered company address is: Gewerbestrasse 11, 6330 Cham, Switzerland

If disposing of this product, please recycle the paper.

Preface

Radio Frequency Identification (RFID)-based backscatter technology boosts battery-free wireless device development and pushes the large-scale deployment of passive Internet of Things (IoT) systems. RFID technology has been paid more ever-increasing attention in a variety of promising applications, such as logistics management, supply chain tracking, environment sensing, gas/liquid leakage monitoring, material identification, health-care monitoring, et al. On account of the ultra-lightweight passive RFID tag and large-scale IoT network deployment, the fundamental problem is to schedule the numerous passive tags to access the network efficiently and ensure the security of data transmission. Driven by these requirements, in this book, we provide a systematic treatment of the theoretical foundations and algorithmic tools necessary in the design and implementation of efficient and secure backscatter networking for RFID applications.

Specifically, focusing on the missing event detection of RFID applications, we deliver a comprehensive treatment on the following problems ranging from theoretically efficient and secure writing scheme and access protocol design to practical system implementation with Commercial Off-The-Shelf (COTS) RFID tags.

- Efficient multiple group labeling scheme in RFID systems.
- Secure anonymous group-wise writing scheme for RFID systems.
- Compact filter-based access protocol for multi-tagged RFID systems.
- Fast and reliable access protocol for multi-tagged RFID systems.
- Practical hashing-free access implementation with COTS RFID systems.

In the book, we unveil a research and exposition line from theoretical modeling and algorithm design to practical COTS RFID system implementation and optimization.

In order to reduce the useless transmission in the RFID system, we start by investigating an efficient multi-seed group labeling scheme in Chap. 2. Specifically, we employ a multi-seed approach to attain efficient group labeling while illuminating the NP-hardness associated with the use of multiple seeds. Due to the NP-hardness of the problem, we introduce an approximate seed assignment algorithm designed to edge closer to the optimal solution. This algorithm selects the slot that is mapped by the highest number of tags within the same group and assigns the corresponding seed to that slot. Moreover, we consolidate the approximation algorithms with a concrete communication mechanism for both the reader and the tags, thereby developing a unified group labeling protocol.

To further improve the security of data transmission in RFID-enabled multi-task backscatter systems, we then come up with a secure, anonymous, group-wise writing scheme in Chap. 3. Specifically, we propose the Overlapped Bloom Filter-based protocol (OBF) and its enhanced version, OBF+. The core is to construct an approximately random sequence as noise by making transmission data for different tag groups overlap with each other, thus hiding the original information with a low computational complexity. The compact filter can guarantee the time efficiency while improving the security of the group writing. To make tags aware of the correctness of the decoded group data, the enhanced version introduces the complementary code-based check mechanism to eliminate the fault data.

We then delve into the issue of missing detection in multi-tagged systems, stemming from the necessity for heightened security and precise object state sensing. Unlike previous studies on single-tagged systems, the response of just one tag attached to an object suffices to confirm the presence of that object instead of all tags in the multi-tagged systems. That said, a pivotal guideline for protocol design is to query a subset of tags rather than the entirety, as advocated in earlier works. Accordingly, we concentrate on two unexplored yet rational avenues for designing missing detection protocols in multi-tagged systems.

Initially, we develop a compact filter-based access protocol for a multi-tagged RFID system to designate and interrogate the tags in Chap. 4. Specifically, we introduce a two-phase Bloom filter-based missing detection protocol, which marks the representative tags (comprising one tag from each object) and facilitates their responses. To enhance the temporal efficiency of missing detection, we replace the Bloom filter with a compressive filter to designate the representative tags and then utilize a composite vector to effectively coordinate their reporting of presence.

Subsequently, we create a fast and reliable access protocol on hash seed searching in Chap. 5. Leveraging the properties of hash functions, an appropriate seed can map one tag from each item to a unique value, thereby enabling the extraction of a subset of tags within the system and assigning them to singleton slots in the response frame for detecting any missing items. The reader first broadcasts the selected seed along with the corresponding unique hashing values, allowing the tags to ascertain when and whether to respond based on the received seed and hashing values. However, the computational complexity of seed searching escalates exponentially with the increasing number of tags. The disparity between the clock frequency of seed searching and the communication bandwidth presents an opportunity to strike a balance between computation and communication, revealing a feasible method for seed searching. Consequently, we design a foundational protocol alongside an enhanced version aimed at further improving time efficiency.

Finally, we set up the practical hashing-free access of the missing event detection platform with COTS RFID systems in Chap. 6. Unlike existing research in this domain, COTS tags lack hashing functionality, rendering them incapable of randomly accessing the reader through hashing mappings. In this context, we establish a theoretical model for missing detection specific to COTS tags and utilize the EPC-global Gen2 standard employed by these systems to devise protocols for missing detection. Specifically, we leverage the Q-command within the Gen2 standard to query the tags, facilitating their random access to the reader and enabling a point-to-multipoint communication pathway from the reader to the tags. To mitigate access collisions in point-to-multipoint scenarios, we subsequently design a point-to-point protocol that is free from collisions by singularizing the tags in each slot with a selective bitmask, thus enhancing the temporal efficiency of missing detection.

Beijing, China
Rongrong Zhang
Hao Liu

Competing Interests The authors have no competing interests to declare that are relevant to the content of this manuscript.

Contents

1 Introduction 1
  1.1 The Universality of RFID Technology 1
  1.2 The Group Writing of RFID Tags 4
  1.3 Missing Event Detection in RFID Systems 6
  1.4 The COTS Implementation of RFID Systems 7
  1.5 Book Organization 7
  References 8

2 Efficient Multiple Group Labeling Scheme in RFID Systems 11
  2.1 Introduction 12
  2.2 Problem Formulation and Motivation 13
      2.2.1 Single-Seed Versus Multi-Seed 14
      2.2.2 Motivation 15
  2.3 Group Labeling Protocol with Multiple Seeds (GLMS) 16
  2.4 Seed Assignment Algorithms 18
      2.4.1 Approximation Algorithm 20
      2.4.2 Simplified Algorithms 24
  2.5 Parameter Configuration 27
  2.6 Performance Evaluation 29
      2.6.1 Simulation Settings 29
      2.6.2 Simulation Results 30
  2.7 Related Work 34
  2.8 Conclusion 36
  2.A Proof of NP-Hardness 37
  2.B Proof of Lemma 1 38
  2.C Proof of Lemma 2 39
  References 39

3 Secure Anonymous Group-Wise Writing Scheme for RFID Systems 41
- 3.1 Introduction 42
- 3.2 Related Work 43
- 3.3 System Model and Problem Formulation 44
  - 3.3.1 System Model 44
  - 3.3.2 Problem Formulation 44
  - 3.3.3 Overview of Our Solutions 45
- 3.4 OBF: Overlapped Bloom Filter-Based Group Writing 46
  - 3.4.1 Motivation 46
  - 3.4.2 Protocol Description 48
  - 3.4.3 Parameters Optimization 49
- 3.5 OBF+: An Enhanced Solution 53
  - 3.5.1 Motivation 53
  - 3.5.2 Protocol Description 53
  - 3.5.3 Parameters Optimization 54
- 3.6 Implementation 59
  - 3.6.1 Experimental Setup 59
  - 3.6.2 Implementation of the Anonymous Group Writing 60
- 3.7 Performance Evaluation 60
- 3.8 Conclusion 70
- References 71

4 Compact Filter-Based Access Protocol for Multi-Tagged RFID Systems ... 73
- 4.1 Introduction 74
- 4.2 Related Work 74
- 4.3 System Model and Problem Formulation 75
  - 4.3.1 System Model 75
  - 4.3.2 Problem Formulation 76
  - 4.3.3 Design Rational 76
- 4.4 Basic Approach: Bloom Filter-Based Protocol 79
  - 4.4.1 Protocol Description 79
  - 4.4.2 Parameter Optimization 80
- 4.5 Advanced Approach: Compressive Filter-Based Protocol 83
  - 4.5.1 Protocol Description 83
  - 4.5.2 Parameter Setting 86
- 4.6 Performance Evaluation 90
- 4.7 Conclusion 96
- References 97

5	**Fast and Reliable Access Protocol for Multi-tagged RFID Systems**	99
	5.1 Introduction	99
	5.2 Related Work	100
	5.3 System Model and Problem Formulation	101
	5.3.1 System Model	102
	5.3.2 Problem Formulation	102
	5.3.3 Design Rational	103
	5.4 M^2ID: Missing Multi-Tagged Item Detection Protocol	105
	5.4.1 Motivation	105
	5.4.2 Segmentation	106
	5.4.3 Protocol Description	107
	5.4.4 Parameter Optimization	108
	5.5 M^2ID+: The Improvement of M^2ID	113
	5.5.1 Motivation	113
	5.5.2 Protocol Description	114
	5.5.3 Parameter Setting	115
	5.6 Performance Evaluation	121
	5.7 Conclusion	124
	References	125
6	**Practical Hashing-Free Access Implementation with COTS RFID Systems**	127
	6.1 Introduction	128
	6.2 Related Work	129
	6.3 System Model and Problem Statement	130
	6.4 P2M: Point-to-Multipoint Missing Tag Identification	131
	6.4.1 Point-to-Multipoint Q-Query	131
	6.4.2 Encoding Methods	132
	6.4.3 Configuration of the Parameter Q	133
	6.4.4 Calculation of the Interrogation Duration	134
	6.5 P2P: Point-to-Point Missing Tag Identification	134
	6.5.1 Point-to-Point Selective Query	135
	6.5.2 Calculation of the Overall P2P Execution Time	136
	6.5.3 *Select* Function	136
	6.5.4 Bitmask Selection	138
	6.5.5 Missing Tag Identification with New Tags	142
	6.6 Implementation	143
	6.6.1 Implementation Setup	143
	6.6.2 Implementation Results	143
	6.7 Conclusion	151
	References	151

7 Conclusion and Perspective 153
7.1 Book Summary 153
7.2 Open Questions and Future Work 154
7.2.1 Energy Utilization 154
7.2.2 Anonymity 156
7.2.3 Tag Implementation 156

Introduction

1.1 The Universality of RFID Technology

Radio Frequency Identification (RFID) technology has become instrumental in the realms of automatic identification, data acquisition, and sensing, thereby earning its distinction as one of the paramount innovations of the twenty-first century [1]. Specifically, RFID technology facilitates the contactless reading and capturing of information from objects by attaching tags on physical items, thus enabling a unique identification for each entity. This capability fosters global object tracking and seamless information dissemination. Moreover, with its proficiency in non-line-of-sight communication and extended range, RFID technology presents a significant advantage over conventional barcode systems [2]. Consequently, RFID technology attracts extensive attention in a variety of applications including industrial manufacturing, logistics management, supply chain tracking [3], environment sensing, material identification, et al. In recent years, with the rapid development of electronic components, the RFID tags are increasingly becoming intelligent and simultaneously diminishing in size and cost, which greatly enhances and broadens the prospects of widespread applications and implementation of the Internet of Things (IoT) [4].

Generally speaking, an RFID system typically consists of one or multiple readers and a vast array of wireless tags. The reader not only functions as a wireless interrogator of tags by receiving data through its antenna but also serves as a conduit for data transmission via a data bus, thereby facilitating both communication and computational functionalities. And the read can adaptively adjust its communication frequency and range. Additionally, it implements an anti-collision protocol to guarantee precise access and identification of the tags. On the other hand, an RFID tag can be integrated with a cost-effective microchip and antenna. And each tag is embedded a unique serial number, i.e., IDentity (ID), stored within its microchip. Based on whether the tag is equipped with an energy source module, the tags can be categorized into three types: active tags, semi-active tags, and passive tags. The active tags, which are equipped with a battery, can periodically monitor wireless communication channels and transmit radio signals. Thanks to the built-in battery, the active tag can work continuously without external supply. However, they are costly and large in size due to the embedded batteries which need to be replaced regularly to maintain their functionalities. Compared to the active tag, the semi-active tag contains a smaller battery which is only sufficient to support the tag to respond when a reader signal is received. Although batteries are still needed for the semi-active tags, they have a longer life and are replaced less frequently. Moreover, they have lower cost and smaller size. Yet, the passive tags are battery free, which draw energy from the reader's signal stimulus to enable the microchip to work and allow the tag to reflect the modulated signal containing its data back to the reader. Without batteries, the passive tags can be designed smaller and less expensive to manufacture. And, the maintenance costs are numerously reduced as no need to replace the battery.

The standardization of RFID systems formulates a series of uniform specifications and guidelines to ensure the compatibility, inter-operability, and efficient application of RFID technology in the world. In recent years, the International Organization for Standardization (ISO) and the International Electrotechnical Commission (IEC) have developed a range of national standards for RFID systems, including the ISO/IEC 18000 series [5]. In particular, the EPC-global standardization emphasizes the integration of RFID systems within the supply chain, with its EPC-global C1G2 [6] becoming the prevalent industrial benchmark for RFID systems. Similarly, Ubiquitous ID espouses a comparable philosophy with EPC-global in the development of RFID system standards, utilizing the ucode encoding scheme. Typically, most researches focused on reader-tag communication conform to the protocols established by EPC-global C1G2.

RFID systems have been deployed in various applications with real-world scenarios, primarily for monitoring objects to detect unauthorized movements, such as theft. In these systems, the readers first interrogate the tags attached to the physical objects within a specific area. Then, the RFID tags respond to the readers via backscatter communication, which can indicate the presence of these physical objects. Correspondingly, an absence of responses manifests a potential missing event. With the deployment of large-scale RFID systems and the increasing requirements of multi-tasking monitoring, each object with multiple tags becomes necessary. However, managing exclusive tags for each task can become impractical and complex. To address this problem, grouping tags that divide RFID tags into multiple groups based on their task requirements is a solution. While this can simplify the tag management, it introduces new challenges in distinguishing tags used for missing event detection from all tags effectively and scheduling numerous tags to access the backscatter network to transmit data. Thus, the design of efficient grouping algorithms and scheduling access protocols is of great importance for missing event detection in RFID systems. This however faces the following challenges:

- *Multi-task tags and multi-tagged objects.* In the field of large-scale warehouse management, countless goods need to be carefully managed, and each good is embedded with tags. With the increasing demands for perception, not only multiple tags are often

attached to each good, but also each tag is used for different tasks. This not only results in an exponential increase in the number of tags, but also makes it necessary to interrogate all tags repeatedly. Therefore, the designed protocols must effectively align the tags with their corresponding tasks and objects to optimize time efficiency.
- *Limited computing and processing capacity on the tag side.* Passive tags are ideally suited for large-scale organization due to their cost-effectiveness and minimal maintenance requirements. The limited energy harvested from stimulus signals signifies that their computational and processing capabilities are constrained, which leads to that the traditional network access protocols are unsuitable for RFID systems. Thereby, the design of protocols must prioritize low-complexity computation and processing.
- *Secure group-wise writing.* Group-wise writing allows the reader to convey its corresponding group data to each tag group. However, due to the sensitivity of group data such as group IDs, there is a risk of eavesdropping when transmitted in plaintext. Traditional well-established encryption algorithms are not suitable for the low-complexity computation and processing demands of backscatter tags. Thus, a lightweight anonymous group writing protocols are urgently needed.
- *Unreliable wireless channel.* The wireless channel between readers and tags is not error-fee due to the channel fading and environmental noise. On account of the limited computing and processing power of tags, traditional channel estimation methods cannot be directly applied to RFID systems. Therefore, the design of protocols must strive to balance the reliability of missing event detection with the inherent uncertainty of the wireless channel.
- *Incompatibility with existing commercial tags.* Hash-based scheduling access protocols have been shown to significantly improve the network efficiency. However, due to the constrained computational capabilities and energy resource limitations of Commercial Off-The-Shelf (COTS) RFID tags, these devices lack the necessary hash functionality. This limitation prevents tags from randomly accessing backscatter networks via hashing mappings. Consequently, in COTS implementation of RFID systems, it is imperative to develop and implement a series of random access network protocols which do not depend on hash mapping, thereby guaranteeing the optimal time efficiency.

These challenges give rise to novel concerns on the design of efficient and secure backscatter networking for missing event detection in large-scale multi-task RFID systems. In this context, this book first introduces a group label scheme that classifies tags into multiple groups, thereby transforming the large-scale system into a composition of several subsystems and thus reducing the access time cost. On the top of this, this book outlines a secure group-wise downlink communication scheme that can enable readers to control tag behavior. Then, the book presents two time-efficient access protocols to improve the missing event detection performance in grouped RFID systems and provides a solution for hashing-free implementation of missing event detection with the COTS RFID tags. Systematically, the book presents state-of-the-art access protocols for missing event detection, highlighting several significant research problems that are both theoretically and practically important. In conclusion, this book tackles a spectrum of issues, ranging from theoretical modeling and analysis to the practical design and optimization of algorithms.

- Efficient group labeling scheme to divide RFID tags into multiple groups, enabling the group-wise query for missing detection.
- Secure anonymous group-wise writing scheme to protect the reader-to-tag downlink control query for missing detection from being eavesdropped.
- Compact filter-based access protocol to schedule tags to transmit data in a time-efficient way guaranteeing the detection accuracy.
- Fast and reliable access protocol to optimize the communication and computation overhead with hash seed searching.
- Practical hashing-free scheduling access implementation with COTS RFID tags.

1.2 The Group Writing of RFID Tags

The group writing serves as a fundamental mechanism for categorizing tags based on usage requirements. For instance, tags affixed to merchandise enable inventory monitoring, while those attached to vehicles in a warehouse function as access credentials for designated operational areas. This group writing approach facilitates multifunctional use of identical tags, thereby significantly reducing management costs. In particular, this book delves into the RFID group writing challenge, which involves accurately and efficiently disseminating category information to all associated tags within an RFID system. Qiao et al. [7] have proposed a polling method for individual tag identification. However, this approach is inefficient due to the redundant transmission of numerous tag IDs OR duplicate group data, allowing only one tag to be labeled per slot. The BIC approach [8] utilizes singleton slots for label assignment but suffers from similar inefficiencies. In contrast, the single-seed protocol CCG [9] leverages slots mapped by the multiple tags within the same group to label several tags simultaneously. Nonetheless, it remains inefficient due to time lost during the transmission of empty slots and those allocated to tags from different groups, particularly when the probability of generating useful slots in a single indicator vector is low. For example, considering 10^3 tags are evenly partitioned into 4, 8, 10 groups, the CCG faces an alarming probability of over 0.6 that a slot fails to label any tags, highlighting a substantial opportunity for enhancement. Consequently, this critical service remains significantly under-explored, presenting substantial opportunities for optimization.

In this book, we firstly introduce a novel group labeling protocol based on a multi-seed framework, enabling multiple mappings from tags to slots. This design allows readers to select the most informative slots from all available mappings, thereby improving the efficiency of data transmission. Specifically, the main context includes the following aspects:

1.2 The Group Writing of RFID Tags

(1) A multi-seed approach is adopted to facilitate efficient group labeling and establish the NP-hardness of the Seed Assignment Problem (SAP) that arises from using multiple seeds for group labeling. This finding elucidates the inherent challenges associated with the group labeling problem, which have not previously been addressed. (2) Given the NP-hardness of the problem, this book proposes an approximate seed assignment algorithm with a competitive ratio of 0.632. This algorithm identifies the slot containing the highest number of tags from the same group at each iteration and assigns the corresponding seed to that slot. Then, leveraging the fact that a tag only receives its associated group data from a single slot, this book develops two simplified algorithms, namely c-search-**I** and c-search-**II**. By capitalizing on the potential for previously useless slots to become useful, these algorithms achieve comparable performance with reduced complexity. (3) Subsequently, this book develops a unified group labeling protocol, named GLMS, which integrates each of the approximation algorithms (AA, c-search-**I**, c-search-**II**) with a well-defined communication mechanism for interactions between the reader and the tags. Additionally, this book investigates optimal parameter configurations to enhance protocol performance.

The group writing scheme allows the reader to concurrently interrogate all tags within the same group to retrieve their group data, such as the group ID. However, transmitting plaintext group data in previous works [8–10] compromises system privacy by exposing sensitive information like the group ID and password, thereby increasing the risk of potential attacks. In this context, there is a need to facilitate anonymous group writing, which effectively informs each tag of its corresponding group data while ensuring the anonymity of this information in the presence of an eavesdropper. The conventional encryption algorithms [11, 12] utilized for securing group data have two significant drawbacks: First, they require integrating a comprehensive encryption/decryption protocol into the original protocol, thereby augmenting communication overhead. Second, the tags must be equipped with the requisite decryption modules, which escalates computational complexity and is not suitable for energy-constrained tags. Consequently, there is an urgent need for a lightweight anonymous group writing protocol that safeguards the privacy of group data in a time-efficient manner.

To this end, this book further introduces an Overlapped Bloom Filter-based (OBF) protocol, along with its enhanced iteration (OBF+), designed to facilitate efficient anonymous group writing through the application of simple logical operators such as OR and AND. The OBF protocol encodes the data of each tag group by applying bit overlapping (logical OR) at positions corresponding to each tag, thereby generating an approximately random sequence that acts as noise on the reader's end. Subsequently, each tag can retrieve the group data from the received bit sequence using logical AND. Building upon the OBF protocol, the OBF+ incorporates the capability to verify the recovered group data, addressing instances of faulty data through the use of data complements. Although augmenting the data with its complement increases the frame size, which may diminish the transmission's time efficiency, this approach effectively eliminates incorrectly recovered group data while simultaneously bolstering the anonymity and overall reliability of the anonymous group writing process.

1.3 Missing Event Detection in RFID Systems

The missing event detection, one of the most widely adopted applications of RFID systems, can significantly reduce financial losses by deploying readers to monitor passive tags attached to products. As RFID systems evolve to meet increasingly complex multi-objective requirements, attaching multiple tags to a single object offers benefits such as enhanced security [13, 14] and precise object state sensing [15–17]. However, this practice introduces challenges related to repeated detection of multi-tagged objects within an expanded system, complicating the swift and reliable identification of missing events. Previous studies [18–27] have not been specifically tailored for multi-tagged RFID systems and suffer from inefficiencies in time management. The primary issue stems from the potential need to detect every tag within the system, which undermines time efficiency in two significant ways: First, current methodologies fail to distinguish between known tags after confirming the presence of one tag on an object, leading to redundant checks and wasted time. Second, the responses of numerous tags on a verified object cause substantial interference with those on other objects. An alternative strategy to avoid repeated presence checks involves selectively polling one tag per object. However, this method necessitates querying each tag with a cumbersome 96-bit ID, which can be time-consuming in large-scale systems. Consequently, the efficient detection of missing events in multi-tagged RFID systems remains an open research question.

Specifically, this book introduces the inaugural formulation and analysis of the missing event detection problem within multi-tagged RFID systems. As aforementioned discussed, a fundamental principle for protocol design is to query a subset of tags rather than the entire collection, as observed in earlier works. Our proposed strategy divides the protocol into two distinct phases: the Marking phase and the Detection phase. During the Marking phase, the reader selects one tag from each object at random and utilizes their mappings to create a filter. This filter effectively identifies the chosen tags, enabling targeted inquiries for further detection in the subsequent phase while suppressing responses from the remaining tags. In the Detection phase, the reader then interrogates the marked tags and identifies missing events based on their responses. Building on this framework, this book will propose two specific two-phase detection protocols: the Basic Protocol and the Advanced Protocol.

Leveraging the properties of hash functions, an appropriate seed that ensures each tag from distinct items maps to a unique value allows for the extraction of a subset of tags within the system. These selected tags are then allocated to singleton slots in the response frame for the purpose of detecting missing items. Subsequently, the reader broadcasts the chosen seed from the initial step along with the corresponding unique hashing values. It is important to note that the disparity between the clock frequency used for seed searching and the communication bandwidth creates an opportunity to establish a trade-off between computation and communication. Thereby, this book has developed two protocols, designated as M^2ID and M^2ID+. The M^2ID protocol outlines the framework for missing event detection, incorporating the computation-communication trade-offs. Building upon this foundation, the M²ID+ protocol aims to further enhance time efficiency.

1.4 The COTS Implementation of RFID Systems

Recently, algorithms for identifying missing tags have garnered significant research attention. However, the current studies on missing tag identification [25, 26, 28–33] often require hashing functionality to facilitate random access to tags, which is not supported by Commercial Off-The-Shelf (COTS) tags. This mismatch poses challenges for practical implementation in real-world scenarios.

Inspired by the aforementioned considerations, this book proposes a comprehensive framework for stable and precise missing tag identification schemes specifically tailored for COTS Gen2 devices. Specifically, we develop two protocols capable of accurately identifying missing tags while remaining fully compatible with the Gen2 standard and existing COTS devices. First, we introduce a point-to-multipoint protocol, referred to as P2M. P2M employs Q-command, which is the de facto random access protocol in the Gen2 standard, to query the tags. This approach effectively accomplishes its task within a bounded worst-case time by meticulously configuring the interrogation duration to 2^Q. To enhance the time efficiency of P2M, this book subsequently designed a point-to-point protocol, known as P2P, which can uniquely identify tags in each slot through the use of a selective bitmask, ensuring reliable communication across all slots. To achieve this, this book further proposes two methodologies for bitmask selection, balancing the trade-off between communication overhead and computational complexity. Ultimately, this book implements both P2M and P2P using COTS RFID devices and rigorously evaluates their performance across various settings.

1.5 Book Organization

In this book, we delineate a pathway that transitions from theoretical modeling and analysis to the practical design and optimization of algorithms for RFID systems. The structure of the book is illustrated in Fig. 1.1. To enhance the reader's experience, we have adopted a modularized framework, wherein each chapter functions as an independent module dedicated to a specific topic as previously outlined. Notably, each chapter includes its own introduction and conclusion sections, which elucidate the relevant work and underscore the significance of the results within the specific context of that chapter. Consequently, we have chosen not to provide an extensive background or comprehensive survey of prior research in this introductory section.

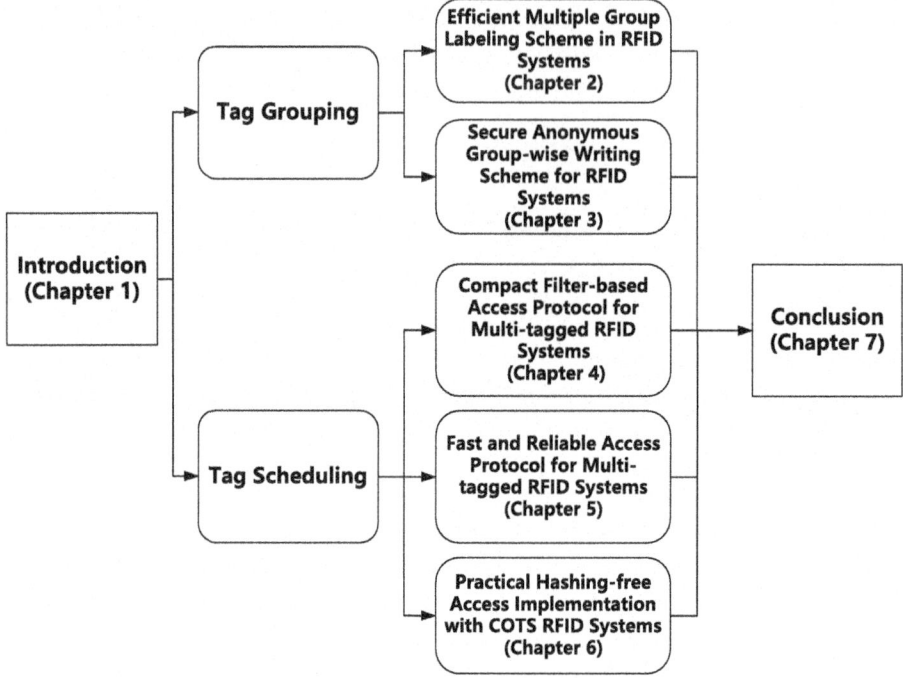

Fig. 1.1 The arrangement of this book

References

1. C. Kaczor, What is an rfid tag? (2024). Available: https://www.camcode.com/blog/what-are-rfid-tags/
2. Barcode (2016). Available: https://en.wikipedia.org/wiki/Barcode
3. SATO Holdings Corporation, Rfid consumables (2019). Available: https://satoasiapacific.com/solutions/rfid/
4. TT Electronics, Rfid: The technology making industries smarter (2022). Available: https://www.ttelectronics.com/blog/rfid-technology/
5. Joint Technical Committee ISO/IEC JTC 1, ISO/IEC 18000-6:2013 (2013). Available: https://www.iso.org/standard/59644.html
6. EPCglobal Inc, Radio-frequency identity protocols class-1 generation-2 UHF RFID protocol for communications at 860 mhz–960 mhz version 1.2.0. [Online]. Available: https://www.gs1.org/sites/default/files/docs/epc/uhfc1g2_1_2_0-standard-20080511.pdf
7. Y. Qiao, S. Chen, T. Li, S. Chen, Energy-efficient polling protocols in rfid systems, in *ACM MobiHoc* (2011), p. 25
8. H. Yue et al, A time-efficient information collection protocol for large-scale rfid systems, in *IEEE INFOCOM* (2012), pp. 2158–2166
9. J. Liu, B. Xiao, S. Chen, F. Zhu, L. Chen, Fast RFID grouping protocols, in *IEEE INFOCOM* (2015), pp. 1948–1956

10. J. Yu, J. Liu, R. Zhang, L. Chen, W. Gong, S. Zhang, Multi-seed group labeling in RFID systems. IEEE Trans. Mobile Comput. **19**(12), 2850–2862 (2019)
11. L. Gao, L. Zhang, M. Ma, Low cost RFID security protocol based on rabin symmetric encryption algorithm. Wireless Personal Commun. **96**(1), 683–696 (2017)
12. J. Yang, B. Liu, H. Yao, Application of chaotic encryption in RFID data transmission security. Int. J. Adv. Network Monitor. Controls **4**(1), 90–96 (2019)
13. L. Bolotnyy, G. Robins, Multi-tag rfid systems. Int. J. Internet Protocol Technol. **2**(3), 218–231 (2007)
14. S. Dhal, I. Sengupta, Protocol to authenticate the objects attached with multiple rfid tags, in *Emerging Trends in Computing and Communication* (Springer, 2014), pp. 149–156
15. L. Shangguan, Z. Yang, A. X. Liu, Z. Zhou, Y. Liu, Relative localization of {RFID} tags using spatial-temporal phase profiling, in *NSDI'15* (2015), pp. 251–263
16. J. Liu, H. Dai, Y. Yan, X. Zhang, X. Chen, L. Chen, Is this side up? detecting upside-down exception with passive rfid, in *IEEE SMARTCOMP* (2017), pp. 1–2
17. D. Hochhalter, D. Bigelow, N. J. Witchey, C. Milam, Rfid-based rack inventory management systems (2018). US Patent App. 15/725,638
18. C. C. Tan, B. Sheng, Q. Li, How to monitor for missing RFID tags, in *IEEE ICDCS* (2008), pp. 295–302
19. W. Luo, S. Chen, T. Li, Y. Qiao, Probabilistic missing-tag detection and energy-time tradeoff in large-scale RFID systems, in *ACM MobiHoc* (2012), pp. 95–104
20. W. Luo, S. Chen, Y. Qiao, T. Li, Missing-tag detection and energy-time tradeoff in large-scale RFID systems with unreliable channels. IEEE/ACM TON **22**(4), 1079–1091 (2014)
21. M. Shahzad, A. X. Liu, Expecting the unexpected: Fast and reliable detection of missing RFID tags in the wild, in *IEEE INFOCOM* (2015), pp. 1939–1947
22. J. Yu, L. Chen, R. Zhang, K. Wang, Finding needles in a haystack: Missing tag detection in large rfid systems. IEEE TCOM **65**(5), 2036–2047 (2017)
23. J. Yu, L. Chen, R. Zhang, K. Wang, On missing tag detection in multiple-group multiple-region rfid systems. IEEE TMC **16**(5), 1371–1381 (2017)
24. J. Yu, W. Gong, J. Liu, L. Chen, K. Wang, R. Zhang, Missing tag identification in cots rfid systems: Bridging the gap between theory and practice, *IEEE TMC* (2018)
25. T. Li, S. Chen, Y. Ling, Identifying the missing tags in a large RFID system, in *ACM MobiHoc* (2010), pp. 1–10
26. R. Zhang, Y. Liu, Y. Zhang, J. Sun, Fast identification of the missing tags in a large RFID system, in *IEEE SECON* (2011), pp. 278–286
27. X. Liu, K. Li, G. Min, Y. Shen, A.X. Liu, W. Qu, Completely pinpointing the missing RFID tags in a time-efficient way. IEEE TC **64**(1), 87–96 (2015)
28. W. Luo, S. Chen, T. Li, S. Chen, Efficient missing tag detection in rfid systems, in *2011 Proceedings IEEE INFOCOM* (IEEE, 2011), pp. 356–360
29. C. Chu, J. Niu, W. Zheng, J. Su, G. Wen, A time-efficient protocol for unknown tag identification in large-scale rfid systems. IEEE IoT J. **9**(15), 13024–13040 (2022)
30. H. Chen, G. Xue, Z. Wang, Efficient and reliable missing tag identification for large-scale rfid systems with unknown tags. IEEE IoT J. **4**(3), 736–748 (2017)
31. M. Shahzad, A.X. Liu, Fast and reliable detection and identification of missing rfid tags in the wild. IEEE/ACM Trans. Netw. **24**(6), 3770–3784 (2016)
32. X. Liu, K. Guo, Z. Liu, X. Zhou, H. Qi, W. Xue, Fast and accurate missing tag detection for multi-category rfid systems, in *2018 IEEE International Conference on Smart Internet of Things (SmartIoT)* (2018), pp. 135–142
33. J. Zhao, W. Li, D.-A. Li, Identifying the missing tags in categorized rfid systems. Int. J. Distributed Sensor Netw. **10**(6), 582951 (2014)

Efficient Multiple Group Labeling Scheme in RFID Systems 2

The group labeling, which involves assigning tags to their respective categories, is not well optimized due to the transmission of redundant data when using only a single seed. In this chapter, we introduce a unified protocol called GLMS (Group Labeling with Multiple Seeds), which employs multiple seeds to construct a Composite Indicator Vector (CIV), thereby reducing useless data transmission. Technically, to address *Seed Assignment Problem* (SAP) that arises during the construction of CIV, we develop an Approximation Algorithm (AA) with a competitive ratio 0.632 by globally searching for the seed that contributes most effectively to the useful slot. We then further design two simplified algorithms through local searching, namely c-search-**I** and its enhanced version c-search-**II**. These algorithms reduce computational complexity by one order of magnitude while achieving comparable performance. Extensive simulations demonstrate the superiority of our approaches.

Chapter roadmap: The remainder of this chapter is organized as follows. Section 2.1 outlines the motivation for studying the multi-seed-based group labeling and summarizes the contributions. The system model, including the problem formulation of the group labeling and the motivation derived from multi-seed hashing, is presented in Sect. 2.2. Section 2.3 details the proposed group labeling protocol utilizing the multiple seeds. In Sect. 2.4, we describe the seed assignment algorithms used to construct the CIV. Section 2.5 investigates how to tune the parameters in the protocol to maximize the time efficiency. Section 2.6 evaluates the performance of proposed approaches compared to state-of-the-art solutions. In Sect. 2.7, we review prior works on group labeling and the existing multi-seed/hash RFID protocols. Finally, we conclude this chapter in Sect. 2.8.

2.1 Introduction

Categorizing the objects (tags) to be monitored into groups is a common practice for efficient management, especially when the system scales (e.g., libraries, supermarkets). A bootstrapping functionality to enable group-wise object management is to inform each object of its group data (e.g., group ID, other related group information), which is named *group labeling*. For example:

- Over-the-air reprogramming on computational RFID tags [1, 2]. These tags work in the same region on a variety of sensing tasks, e.g., temperature, humidity monitoring, and intrusion detection. We regard the tags carrying out the same mission as belonging to the same group. In such a scenario, it is necessary to maintain and upgrade the firmware of tags wirelessly. Since the firmware for tags in different groups is usually different, the system administrator must reprogram categorized tags correctly. That is to say, data for one group should not be received by tags in the other groups.
- Group ID-enabled applications. When the administrator needs to frequently check the status of the expiry-date-sensitive objects, grouping the objects (tags) with the similar expiry date is necessary, wherein group IDs play an important role. Specifically, if the tags with similar expiry dates share the same group ID, the reader can send the required data together with the group ID once to all group members, which not only sharply reduces the communication cost in comparison with the traditional unicast transmission, but also is prerequisite of diverse queries in RFID systems, such as tag estimation [3, 4], top-k query [5, 6] and missing tag detection [7].

While due to the nature of RFID, a tag has neither information about the other tags nor its group, it thus does not know which data is only for its group. In this context, group labeling is called for to correctly tell each tag the data for its group and facilitate the tag management illustrated above.

This chapter presents a multi-seed-based protocol enabling multiple mappings from tags to slots so that the reader can pick up the most informative slots among all mappings for the data transmission and the efficiency is thus improved. The key challenge lies in how to find these slots while achieving seed assignments with low complexity. The superiority and novelty of our method compared with the existing ones are four-fold:

1. Empty slots and those mapped by multiple tags from different groups under one seed which are wasted in [8], can be used to label tags with another seed in our method.
2. The impact of multiple mappings on-tag collisions of different groups is weakened. With different seeds a tag mapped to multiple slots actually receives its group data only in one slot and will keep silent, reducing the collision probability of different groups in the subsequent slots.

3. Collision slots with tags from the same groups instead of only singleton slots or empty slots in the existing work [9, 10] are exploited in our method, improving time efficiency. Moreover, a k-good slot that can label k tags of the same group, can become k^+-good where k and k^+ are constant and $k^+ > k$, significantly reducing the labeling delay.
4. This chapter is the first work formally proving the NP-hardness of the formulated problem arising from the application of multiple seeds and designing the approximation algorithms for the group labeling problem, which makes the mathematical nature of our work completely different from the existing ones and more challenging.

Therefore, we first use a multi-seed approach to achieve efficient group labeling in which we find NP-hardness of the Seed Assignment Problem (SAP) arising from the employment of multiple seeds. To address this issue, we propose a suboptimal solution that selects the slot with the most tags from the same group each time among all slots and assigns the corresponding seed to this slot. Then, we develop another two simplified algorithms, namely c-search-I and c-search-II via converting the originally useless slots to useful. To consolidate each approximation algorithm (AA), c-search-I, c-search-II with a concrete communication mechanism for the reader and tags, we develop a unified group labeling protocol named GLMS.

Our multi-seed protocol generalizes the existing single-seed protocols with remarkably better performance. Our test results show that GLMS achieves a gain of up to 34.2% in terms of the group labeling time.

2.2 Problem Formulation and Motivation

We study an RFID system of one or multiple readers and a number of tags, wherein the tags are partitioned into multiple groups and the readers are connected via high-speed channels with a back-end server of powerful computing capability. We regard the server and the reader(s) as a single entity called *the reader* for simplicity [10, 11]. Generally, the tags have user-defined memory to achieve the writing and storage of the user-defined data [12]. Moreover, we assume that the reader has the IDs of all tags in the system, commonly in designing application-oriented protocols, e.g., missing tag event detection [10, 13] and information collection [9, 14]. To streamline the presentation, we first consider the single-reader case and discuss the multi-reader case later.

Consider a set $X = \{x_1, x_2, \cdots, x_N\}$ of N tags whose IDs are recorded in the reader divided into G disjoint groups. Suppose the size of group g ($1 \leq g \leq G$) is N_g and we have $\sum_{g=1}^{G} N_g = N$. We denote by d_g the data for group g ($1 \leq g \leq G$). In this chapter, we are interested in addressing the following problem: *The group labeling problem is to devise a protocol to send each group data correctly to all its members (tags) within the minimum time.* By correctly, we mean that the data for one group should not be received by tags of the

Table 2.1 Main notations

Symbols	Descriptions
k-good	Useful slot with k tags
N	The number of tags in the system
G	The number of groups
g, d_g	Group index, data for group g
N_g	The number of tags of group g
f, l	Frame size, the number of seeds
s_i	The i-th seed
C_{ij}	The set of tags mapped to j-th slot under s_i
m	The number of labeled tags in the current round
z	The number of chosen useful slots in the current round
u	Time efficiency
N'	Unlabeled tags in current round
N'_g	Unlabeled tags of group g in the current round
G'	The number of groups with unlabeled tags
$\overline{f}$	Upper bound of f
$\overline{l}$	Upper bound of l

other groups. The performance metric is the communication cost between the reader and the tags. Table 2.1 summarizes the main notations used in the chapter.

2.2.1 Single-Seed Versus Multi-Seed

The communication between the reader and tags follows the frame-slotted Aloha protocol [15]: the reader initiates communication first by broadcasting commands containing the parameters, such as frame size f, l random seed(s) s_i with $i \leq l$. In the existing single-seed protocols where $l = 1$, each tag uses its ID and the received seed to generate **one** pseudo-random number via hash function $H(ID, s_1)$ and then maps itself to the slot ($H(ID, s_1)$ mod f) in the frame. On the contrary, in our multi-seed protocol where $l \geq 1$, each tag holds **multiple** pseudo-random numbers with l different seeds and is mapped to l slots in the frame and the most useful slot will be chosen by the reader to send data as introduced shortly.

In this chapter, we make the following definitions on slot states: 1. *Empty slot*: Consider an arbitrary slot, if no tag is mapped to this slot; 2. *Heterogeneous slot*: if multiple tags from different groups are mapped to this slot. 3. *Useless slot*: if this slot is either empty or heterogeneous. If the reader sends data in such a slot, either no tag receives data or tags

2.2 Problem Formulation and Motivation

from one group receive data of another group, which should be avoided in the group labeling problem; 4. *Useful slot*: if tag(s) from the same group is mapped to this slot. In such a slot, the reader can send data to tag(s) from the same group. 5. *Reparable slot*: A slot is reparable if it becomes useful from a heterogeneous slot as the protocol runs, which will happen when tag(s) blend with the others from another group and stay silent after being assigned useful slots.

2.2.2 Motivation

As an indicator vector constructed from a single mapping generates limited useful slots, much time is wasted on the transmission in the useless slots. If multiple seeds are used to generate multiple mappings, the reader can pick up the most informative slots from them to build a composite indicator vector (CIV), reducing the number of the useless slots. Intuitively, assume a slot in a single indicator vector is useful with the probability of 0.5, then with l seeds used to map the tags this probability is $1 - (1 - 0.5)^l$, which quickly approaches 100% with the increase of l.

In addition to increase the number of useful slots, using multiple seeds can also contribute to more labeled tags. Let k-good define a useful slot with k tags. A slot may be k-good under one seed but k^+-good under other seeds where $k^+ > k$, which can be interpreted from the following toy example.

Example 1. Consider an RFID system with two tag groups $G_1 = \{x_1, x_2\}$ and $G_2 = \{x_3, x_4, x_5\}$ and suppose a frame of four slots and two seeds s_1, s_2. From Fig. 2.1 where the shaded rectangles stand for the useful slots, we find just partial slots useful after either mapping, but a CIV of all slots being useful can be built by selecting the most informative slots from two mappings. Specifically, by designating s_1 for the first and third slots, and s_2 for the second and fourth slots, we can build a CIV indicating the seed assignment for each

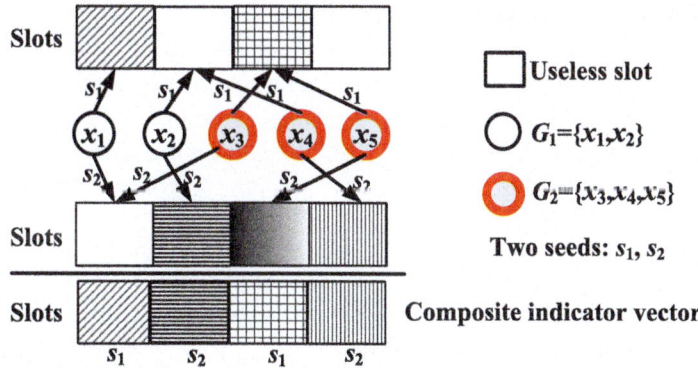

Fig. 2.1 Exemplifying the motivation: the shaded rectangles typify useful slots

slot so that all slots to be executed become useless ones (e.g., the 2nd slot under s_1) to useful ones (e.g., the 2nd slot under s_2) and from 1-good one (e.g., the 3rd slot under s_2) to 2-good one (e.g., the 3rd slot under s_1).

Motivated by the above observation, we design a series of seed assignment algorithms to build the CIV, and develop a unified group labeling protocol, named GLMS, to consolidate each algorithm with the concrete communication mechanism for the reader and tags, respectively. Note that the designed seed assignment algorithms are used in the first phase of the group labeling protocol GLMS. In the following, we first introduce the group labeling protocol and elaborate on how to build the CIV, subsequently.

2.3 Group Labeling Protocol with Multiple Seeds (GLMS)

The execution of the protocol GLMS consists of multiple rounds, each having three phases referred to as *initialization phase*, *screening phase*, and *labeling phase*, respectively. The reader first uses one of the seed assignment algorithms, namely AA, c-search-I, and c-search-II, to be introduced in Sect. 2.4 to build a CIV that determines a unique tag-seed-slot relationship. In the screening phase, the reader sends the CIV to inform each active tag of whether and when it is scheduled to receive its associated group data. In the labeling phase, the reader transmits group data in the designated slots to the eligible tags. If a tag receives its associated group data, it will keep silent in the subsequent rounds. The process of GLMS and the core function of each phase are illustrated in Fig. 2.2.

Protocol Description. Consider an arbitrary round in the execution of the protocol GLMS. Let N', N'_g denote the number of the remaining overall unlabeled tags and that of unlabeled tags of group g at the beginning of this round, respectively. And denote by G' the number of the groups with unlabeled tags. If it is the first round, it holds that $N = N'$ and $G = G'$. The l seeds denoted as s_i, $1 \leq i \leq l$, are used in this round to generate the CIV of f slots. Our multi-seed protocol GLMS is shown in Algorithms 1 and 2.

(1) Initialization Phase: Given l seeds and the frame size f, the CIV can be compounded from l mappings, each involving a different seed. How the values of f and l are chosen will be analyzed in Sect. 2.5 on the parameter optimization. Specifically, in the i-th mapping, we employ seed s_i to map each active tag to one of f slots in the frame. With all l seeds used, the reader records l vectors, each consisting of f cells storing tags mapped to the corresponding

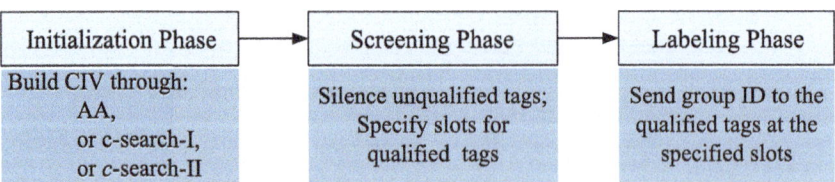

Fig. 2.2 The process of GLMS: Initialization phase, screening phase, and labeling phase in sequence

2.3 Group Labeling Protocol with Multiple Seeds (GLMS)

slots. Using one of the seed assignment algorithms introduced in Sect. 2.4, the reader can designate one seed for each slot in the CIV maximizing the time efficiency in this round.

More specifically, based on the seed assignment, the reader builds a CIV of f slots each of which corresponds to a slot in the frame at the same position and stores the index of the assigned seed. If designating seed s_i for a slot j, the reader stores i that is the index of s_i in the j-th slot of the CIV. If a slot is still useless after l mappings, the reader sets its value in the CIV to zero. Consequently, the positions of non-zero value in the CIV stands for the useful slots of the frame. As there are l seeds, we need $\lceil \log_2(l+1) \rceil$ bits to record one seed's index, that is to say, the length of the CIV is $f \cdot \lceil \log_2(l+1) \rceil$.

Note that if a tag is mapped to a useful slot as specified in the CIV, we refer to this slot as *the useful slot for this tag*.

(2) Screening Phase: The reader broadcasts a message containing the built CIV, the frame size f and l seeds $s_1, s_2, \cdots, s_l$. Upon receiving the message, each tag can extract two pieces of information from the CIV: One is whether the tag is eligible to receive its group data in this round. Specifically, each tag can employ the received l seeds to select l slots in the frame and knows the corresponding l positions it is mapped to in the CIV. Based on the rule of generating the CIV, if a tag is mapped to the j-th position in the CIV under seed s_i and the value in that position is i, then the tag regards slot j as the useful slot for it. In case the conditions can be satisfied under multiple seeds, the tag only selects the slot with the smallest value of j. While if a tag fails under all seeds, it does not participate in any activity until the next round.

The other one is which slot a qualified tag should actually wait for its group data. Because the CIV may contain zero elements which stand for the useless slots, the reader needs to remove the corresponding slots before starting the frame to transmit group data for saving time. The key here is that the tag must know which slots are removed. To that end, we use the ordering approach [14]. Assume slot j is the useful one for the tag, the tag first checks every position before the position j in the CIV. If there exist $\hat{j}$ non-zero elements, the tag will select $(\hat{j}+1)$-th slot to receive its group data and $\hat{j} < j$.

Let us see an example shown in Fig. 2.3. Consider an arbitrary tag x_9. With seeds s_1, s_2, s_3, x_9 is mapped to the 2nd, 1st and 4th slots. After checking the corresponding positions in the CIV, x_9 finds only the 4th element equal to 3 which is the index of s_3, so it regards slot 4 as its useful slot. Furthermore, as there exist two non-zero elements before the 4th position in the

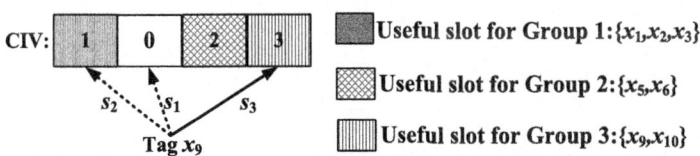

Fig. 2.3 Interpreting indicator vector: s_1, s_2, s_3 are the used seeds

---
Algorithm 1: GLMS for the reader
---
1 // Phase one - the CIV construction
2 Generate l seeds $s_1, s_2, ..., s_l$ randomly
3 Map the unlabeled tags into l slots and generate $f \times l$ cells each recording a set of the tags mapped to the corresponding slot
4 Build CIV via AA or $c-$search-I or $c-$search-II; record the number of non-zero slots in the CIV z
5 // Phase two - CIV transmission
6 Issue a frame start command, transmit CIV and the corresponding frame size f and l seeds
7 // Phase three - tags labeling
8 **for** $i = 1$ to z **do**
9   Issue slot-start command
10   Broadcast the corresponding group ID to the tags mapped to the i-th homogeneous slot
11 **end**
12 Update the set of the unlabeled tags and initiate the next round
---

CIV, x_9 will wait for its group data at slot 3 in the labeling phase. Therefore, only the three useful slots will be executed in the labeling phase instead of the four in the original frame.

(3) Labeling Phase: After the qualification test in the screening phase, only the eligible tags partake in this phase. By knowing all tag IDs and the CIV, the reader knows the order of the slots actually selected by the eligible tags. Assume there are z non-zero positions in the CIV, the reader initiates a labeling frame of z slots and sends the corresponding group data at each slot to the eligible tag(s) for which this slot is useful. As the tag(s) in each slot come from the same group, they can be labeled simultaneously. On the other hand, each tag learns from the CIV at which slot the reader will transmit its group data and can thus receive the data at that slot.

For instance, recall the example in Fig. 2.3, the reader actually initiates a frame containing three useful slots in Fig. 2.3. It can label tags x_1, x_2, x_3 by sending ID of group 1 in the slot 1, and label tags x_5, x_6 and x_9, x_{10} in the slots 2 and 3, respectively.

After the current round, the reader moves to the next round, which is identical except that the labeled tags will keep silent. That is, only the unlabeled tags attend the next round. The above process repeats round after round until all tags receive their associated group data.

In what follows, we start formally presenting the seed assignment algorithms used to build the CIV.

2.4 Seed Assignment Algorithms

The key to our multi-seed method lies in the seed assignment arising in building the CIV. Specifically, given l seeds s_i ($1 \leq i \leq l$) and the frame size f, the reader needs to designate one seed for each slot in the CIV and inform each tag of the seed assignment by sending the CIV. Therefore, if the CIV is built the tags mapped to each slot are deterministic.

2.4 Seed Assignment Algorithms

Algorithm 2: GLMS for tags

1. Receive the CIV and the corresponding frame size f and l random seeds
2. Compute l mapped slot number $sn[i] = H(f, ID, s_i)$
3. Initialize the current slot number $csn \leftarrow 1$ and current random seed index $ci \leftarrow 0$
4. **while** *TRUE* **do**
5.   Wait-for-slot-start().
6. $j \leftarrow$ the number of zeros in the first csn positions in CIV
7. $ci \leftarrow CIV[csn + j]$
8.   **if** $(csn + j) == sn[ci]$ **then**
9.   Store the received Group ID.
10.   **end**
11. $csn \leftarrow csn + 1$
12. **end**

More specifically, recall that the CIV of f slots is compounded from l mappings, there are $l \times f$ cells in total each of which records a set of the tags mapped to the corresponding slot, as shown in Fig. 2.4. C_{ij} stands for the set of the tags mapped to slot j under seed s_i for $1 \le i \le l$ and $1 \le j \le f$, and $1 \le I_j \le l$ denotes the index of the seed finally assigned for slot j in the CIV, and C_j is the set of tags that will be mapped to slot j under seed s_{I_j} following the built CIV. Note that since l seeds are used and zero represents useless slots, we need $\lceil \log(l + 1) \rceil$ bits to stand for each seed index I_j. Moreover, it may happen that $I_j = I_{j'}$ for $j \ne j'$ because a seed may be assigned to multiple slots in the CIV.

As a tag can be mapped to l positions under l different seeds, slots from multiple mappings may share the same tags, that is, $C_{ij} \cap C_{i'j'} \ne \emptyset$ for $i' \ne i$ and $j' \ne j$. Define a set comprising tags from the same group as a pure set which is equivalent to a useful slot. Define the time efficiency u as the number of tags labeled per unit time. Recall that if a seed is designated for a slot, then the tags mapped to this slot under this seed are deterministic. In this sense, we should carefully assign seeds such that the time efficiency u can be maximized.

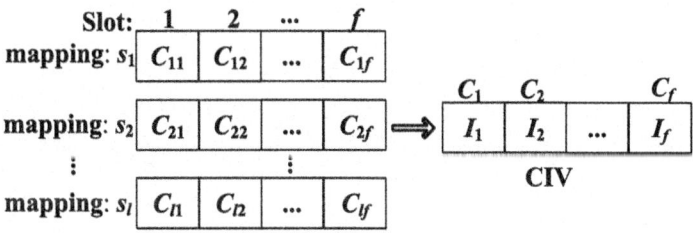

Fig. 2.4 Exemplifying the seed assignment problem: C_{ij} is the set of the tags mapped to slot j under seed s_i; I_j denotes the index of the seed assigned for slot j in the CIV; C_j is the set of the tags mapped to slot j under seed s_{I_j}

Given a seed assignment, let z be the number of yielded useful slots and let $m = |\cup_j C_j|$ be the size of the union of the tags mapped to the useful slots. Let t_0 and t_g denote the time for the reader to transmit one bit and data for group g, respectively. Without loss of generality, we assume the data size for each group is identical. With (2.2), we formally define the following seed assignment problem.

Problem 1 (Seed assignment problem) Given $l \times f$ sets of the tags C_{ij} for $1 \leq i \leq l$ and $1 \leq j \leq f$, and define S as the collection of the seeds assigned to each slot in the CIV, the seed assignment problem is to seek S satisfying

$$S = \underset{s_{l_j}}{\mathrm{argmax}} \frac{|\cup_j C_j|}{t_g(a+z)},$$

where $a = f \lceil \log_2(l+1) \rceil t_0/t_g$. That is to say, given the seeds and the frame size, the reader seeks an optimum collection S of the seeds which will maximize the time efficiency u.

Problem 1 performs combinatorially, which is usually NP-hard. The challenge here lies in how to prove its NP-hardness. In the following, we formally state the NP-hard observation and its proof.

Theorem 1 *Problem 1 is NP-hard.*

Proof For clarity, we just outline the proof here and the complete proof is provided in Appendix 2.A. To study the hardness of Problem 1, we prove it polynomially reducible from the Maximum coverage problem [16] which is a classic NP-hard problem. Given h sets and an integer $k \leq h$ with which we need to solve the Maximum coverage problem, the polynomial reduction comprises three steps: First, we replicate each set k times and obtain $h \times k$ sets. Second, we introduce k dummy sets to guarantee that each slot in the CIV is assigned only one seed. Third, we prove that u reaches its maximum only when k sets are chosen in Problem 1. □

Due to the NP-hardness of SAP, in what follows, we design a series of algorithms to approach the optimal time efficiency. Specifically, we first design an approximation algorithm (AA) and develop two simplified algorithms with the less complexity but good performance on the top of AA.

2.4.1 Approximation Algorithm

Motivation
Recall the Problem 1 that seeks the seed assignment to maximize time efficiency u, we can achieve this objective from two directions. On the one hand, we want to use fewer useful

2.4 Seed Assignment Algorithms

Algorithm 3: Approximation algorithm for Problem 1

Input : s_i, f
Output: u_{max}, tags in picked slots C, seed assignment S
1 **Initialisation:** $C, S \leftarrow \emptyset; R, z, u_{max} \leftarrow 0; H \leftarrow (C_{ij})_{l \times f}$
2 **while** $j_1 \leq f$ **do**
3   // Search the most useful slot
4   **for** $j = 1$ *to* f **do**
5   **for** $i = 1$ *to* l **do**
6   **if** C_{ij} *is useful and* $|C_{ij}| > R$ **then**
7              $R \leftarrow |C_{ij}|, I \leftarrow i, J \leftarrow j$
8   **end**
9   **end**
10   **end**
11   // Select the seed contributing to the most useful slot
12   **if** $\frac{|C \cup C_{IJ}|}{t_g(a+z+1)} \geq u_{max}$ **then**
13        $S \leftarrow S \cup (s_I, J)$ /* Assign seed s_I to slot J */ $C \leftarrow C \cup C_{IJ}$, and $z \leftarrow z+1$
14        $u_{max} \leftarrow \frac{|C \cup C_{ij}|}{t_g(a+z)}$
15   **else**
16   Stop
17   **end**
18   // Clear the slots at J-th column in Fig. 2.4 and deduct the tags in the picked slot from the remaining slots
19   **for** $j = 1$ *to* f **do**
20   **for** $i = 1$ *to* l **do**
21   **if** $j == J$ **then**
22              $H \leftarrow H/C_{ij}, C_{ij} \leftarrow \emptyset$
23   **else**
24              $C_{ij} \leftarrow C_{ij} - C_{IJ}$
25   **end**
26   **end**
27   **end**
28   **if** $H == \emptyset$ **then**
29   Stop
30   **end**
31 **end**
32 Return u_{max}, C, S

---

slots, i.e., minimizing z, while maximizing the number of the tags m involved in these used useful slots. Observing the waste of heterogeneous slots (c.f. Sect. 2.2.2) in the prior work, we, on the other hand, hope to design an algorithm that is able to exploit the heterogeneous slots that can become useful as the algorithm runs.

Overview

Define the most useful slot as the useful slot with the most tags from the same group. The core idea of AA lies in that each time the reader selects the seed contributing to the most useful slot to maximize the time efficiency u. Note that there is a unique seed-tag-slot mapping, that is, given any two of them, we can fix the third one. Since a set of the tags (c.f. Fig. 2.4) is indexed by the used seed and the mapped slot, once a most useful slot is found the reader assigns the corresponding seed to this slot and knows the tags mapped to this slot, which are referred to as covered tags here.

Moreover, to enable the utility of heterogeneous slots, the reader first deducts the covered tags from the remaining non-empty slots including both heterogeneous and useful slots, and then checks their states and picks the most useful one among them. The rationale behind this is that each covered tag will stay silent after its corresponding most useful slot so that actually it will not be blent with tags in the subsequent slots under all mappings, which enables the conversion of a subsequent heterogeneous slot into a useful one. Note that we refer to such a heterogeneous slot as *reparable slot*.

Algorithm Description

Formally, we illustrate the AA in Algorithm 3 with the input of l seeds and the frame size f. It is easy to check that the computation complexity of AA is $O(l \cdot f^2)$. The main procedures of AA are summarized below.

- Each time the reader

    - picks the most useful slot to which the most uncovered tags are mapped and brings the most gain in time efficiency u. (Line 4-12 in Algorithm 3)
    - records the subscripts of the chosen slot standing for which seed will be assigned to this slot. (Line 13)
    - records the tags in the chosen slot, marks them as covered, and removes them from the remaining slots. Since only one seed should be assigned to each slot in the CIV, the slots under the other mappings but in the same column (c.f. Fig. 2.4) as the chosen most useful slot would be emptied. (Line 19-27)

- The algorithm stops if there is no useful slot or no useful slot contributing to the greater time efficiency.
- The algorithm outputs the seed allocation for each slot in the CIV and a collection of the covered tags, with which the time efficiency u is maximized under the given input.

After executing Algorithm 3, the reader builds a CIV and knows which tags can be labeled in which slots. Specifically, if a set C_{ij} in the useful slot is chosen, then the reader designates

2.4 Seed Assignment Algorithms

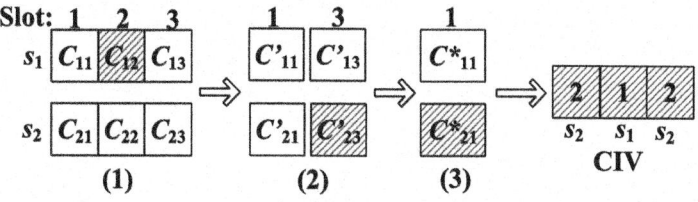

Fig. 2.5 AA: the streak represents the unselected most useful slot

---
Algorithm 4: c-search-**I** for Problem 1

Input : s_i, f, c
Output: u_{max}, tags in picked slots C, seed assignment S
1 Initialisation: $C, S \leftarrow \emptyset; R, z, u_{max} \leftarrow 0; H \leftarrow (C_{ij})_{l \times f}$
2 **while** $j_1 \leq f$ **do**
3 $\quad$ Choose c columns out of unselected ones randomly
4 $\quad$ // Search the most useful slot from the c columns: define $j_{j'}$ as the j'-th chosen column
5 $\quad$ **for** $j' = 1$ to c **do**
6 $\quad\quad$ **for** $i = 1$ to l **do**
7 $\quad\quad\quad$ $C_{ij_{j'}} \leftarrow C_{ij_{j'}} - C_{IJ}, H \leftarrow H/C_{iJ}$
8 $\quad\quad\quad$ **if** $C_{ij_{j'}}$ is useful and $|C_{ij_{j'}}| > R$ **then**
9 $\quad\quad\quad\quad$ $R \leftarrow |C_{ij_{j'}}|, I \leftarrow i, J \leftarrow j_{j'}$
10 $\quad\quad\quad$ **end**
11 $\quad\quad$ **end**
12 $\quad$ **end**
13 $\quad$ Conduct the operations as lines 12 − 14 in Alg. 3
14 **end**
15 Return u_{max}, C, S
---

seed s_i for the slot j and sets the value of the slot j in the CIV to i. In case that all sets in the column j in Fig. 2.4 are not chosen, the reader sets the slot j's value to zero in the CIV.

Next, we illustrate AA in Fig. 2.5 with 2 seeds and a frame of 3 slots. First, the reader finds C_{12} the most useful, then it assigns s_1 to slot 2 in the CIV and empties C_{i2} while removing the tags in the intersections between C_{12} and the others, yielding C'_{ij}. Repeating the operations, the reader finds C'_{23} the most useful via searching from the columns 1 and 3, and then C^*_{21} from the columns 1 in sequence. Finally, the reader builds the CIV as shown in Fig. 2.5. To evaluate algorithm performance, we derive the competitive ratio of the algorithm.

Lemma 1 (*Competitive ratio of Algorithm 3*) *Let u_{opt} denote the optimal time efficiency of Problem 1, it holds for the time efficiency u_{max} of Algorithm 3 that $u_{max} \geq 0.632 u_{opt}$.*

Proof The proof is detailed in Appendix 2.B. $\square$

2.4.2 Simplified Algorithms

For better scalability to the system scale, we here present two simplified algorithms, namely c-search-**I** and its improved version: c-search-**II**, to reduce the complexity of AA while achieving the comparable performance.

c-**search-I**

The key difference of c-search-**I** from AA consists in locally searching the most useful slot among the c columns in Fig. 2.4 chosen **randomly** each time instead of global searching among all f columns in AA. At first glance, this simplified operation would degrade the performance significantly, but besides the less complexity, another advantage of this is curing more heterogeneous slots, which benefits to the increase in time efficiency. Look at an example with the frame size f and $c \leq f$. Assume that the first most useful slot in AA occurs at one of the mappings in f-th column of Fig. 2.4, then none of the heterogeneous slots can become useful. This is because a tag mapped to a heterogeneous slot can be eliminated from this slot only when this heterogeneous slot is later than the most useful slot for this tag. While in c-search-**I**, if we find the first most useful slot in $f/2$-th column by locally searching among c randomly chosen columns, then we can exploit the subsequent reparable slots.

Algorithm 5: c-search-**II** at $c = 1$ for Problem 1

1 **while** $j_1 \leq f$ **do**
2   // Search the most useful slot from the j_1-th column
3   **for** $i = 1$ *to* l **do**
4        $C_{i,j_1} \leftarrow C_{i,j_1} - C_{IJ}, H \leftarrow H/C_{IJ}$
5   **if** C_{i,j_1} *is useless* **then**
6           $C_{i,j_1} \leftarrow \varnothing$
7   **else if** $|C_{i,j_1}| > R$ **then**
8           $R \leftarrow |C_{i,j_1}|, I \leftarrow i, J \leftarrow j_1$
9   **end**
10   **end**
11   The remaining steps are the same as c-search-**I**
12 **end**

We list c-search-**I** in Algorithm 4 with a new input c and summarize the main procedures as below: Each time the reader

- chooses c columns from unselected ones randomly, containing $c \cdot l$ slots.
- removes the covered tags from these chosen slots.
- picks the most useful slot among the slimmed-down $c \cdot l$ slots, which achieves the most gain in time efficiency u.

2.4 Seed Assignment Algorithms

- records the subscripts of the chosen slot standing for which seed will be assigned to which slot.
- records the tags in the most useful slot picked and marks them as covered.

Next, we illustrate the influence of c on the performance.

Example 2. In the experiment, we partition 1000 tags evenly into $G = 2, 4, 8, 10$ groups and vary c from 1 to 40. Figure 2.6a shows that the time overhead at $c = 40$ is the least, which is very close to AA. For the tradeoff between the complexity and performance, we will set $c = 40$ in the simulation in Sect. 2.6.

c-search-II

As described above, c-search-I achieves the comparable performance with the less complexity, but it may fail to exploit the reparable heterogeneous slots furthest. For example, if the first most useful slot in c-search-I arises in $f/2$-th column among c randomly chosen columns, then we cannot exploit the potential reparable slots in the first $(\frac{f}{2} - 1)$ columns. To address the issue in c-search-I, we propose an improved algorithm, named c-search-II, pursuing less complexity but better performance than c-search-I.

The main difference from c-search-I is that c-search-II chooses c columns among the unselected columns *in the ascending order of the column number* instead of randomly. For instance, assume $c = 10$, we choose the columns 1–10 as the candidates (c.f. Fig. 2.4). In the case that columns 1, 3 and 4 have been chosen previously, we will select columns 2 and 5–13. Next, we would like to take an example to explain the main differences among AA, c-search-I and c-search-II.

Example 3. We show the first round operation of the three algorithms in Fig. 2.7 where we suppose $c = 2$ in two simplified algorithms. Specifically, AA finds C_{24} the most useful slot

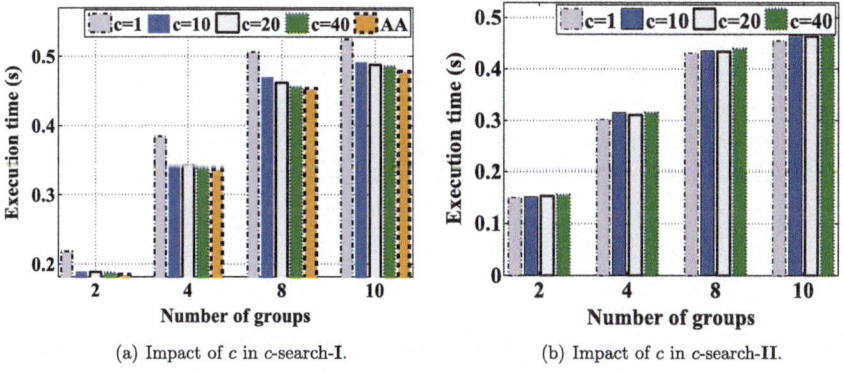

(a) Impact of c in c-search-I. (b) Impact of c in c-search-II.

Fig. 2.6 Impact of c in c-search-I and c-search-II

Algorithm 6: Seeking the optimal f and l

Input : $N', G', step, t_1$ and t_{gid}
Output: u^*, f^*, l^*
1 **Initialisation:** $\overline{f} = \infty, \overline{l} = \infty, u^* = 0, Q = 0$
2 **while** $f \leq \overline{f}$ and $l \leq \overline{l}$ **do**
3   Execute Algorithms 3 or 4 or c-search-**II**
4     $u = u_{max}$ returned from the executed algorithm
5     $Q = Q + 1$, find f_{q^*}, l_{q^*} with $\underset{1 \leq q \leq Q}{\mathrm{argmax}}\, u(f_q, l_q)$
6     $f^* = f_{q^*}, l^* = l_{q^*}, u^* = u(f_{q^*}, l_{q^*})$
7   Update $\overline{f}$ with (2.3) and update $\overline{l}$ with (2.4)
8     $f = f + step, l = l + 1$
9   **if** $f > \overline{f}$ **then**
10   **if** $l \leq \overline{l}$ **then**
11           $f = 1 : step : \overline{f}$
12   **else**
13   Stop
14   **end**
15   **else if** $l > \overline{l}$ **then**
16       $l = 1 : \overline{l}$
17   **end**
18 **end**
19 Return optimum efficiency u^* and the optimum (f^*, l^*)

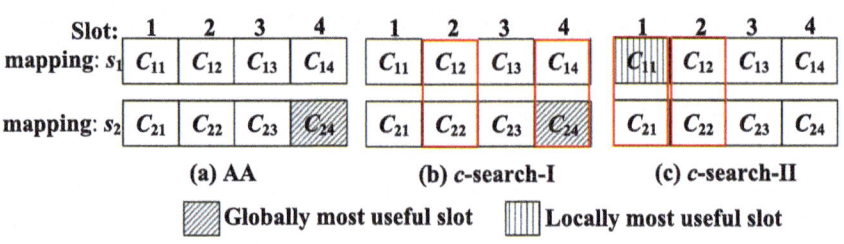

Fig. 2.7 Difference among three algorithms: s_1 and s_2 are two seeds

by globally searching among $2 * 4$ cells, while c-search-**I** first selects two columns randomly (assume that columns 2 and 4 are chosen), and searches for the most useful slot among $2 * 2$ cells. Differently, c-search-**II** chooses the first two columns and then searches among the corresponding $2 * 2$ cells. As C_{11} is found the most useful in c-search-**II**, the reparable slots in the columns 2–4 can be exploited later.

In this chapter, *we will set c to 1 in c-search-II* and state the seed assignment process in Algorithm 5. The rationale behind the setting is that with $c = 1$ we can employ the potential reparable slots to the greatest extent, namely those in the columns $2-f$. Besides, the yielded complexity is $O(l \cdot f)$ which is less than $O(c \cdot l \cdot f)$ in c-search-I and $O(l \cdot f^2)$ in AA,

2.5 Parameter Configuration

Table 2.2 Algorithm complexity with l seeds and the frame size f

Algorithm	AA	c-search-**I**	c-search-**II**
Complexity	$O(l \cdot f^2)$	$O(c \cdot l \cdot f)$	$O(l \cdot f)$

which are listed in Table 2.2. Under the settings as in Example 2, we show in Fig. 2.6b that c-search-**II** achieves good performance at $c = 1$. We will further evaluate the performance of c-search-**II** at $c = 1$ in Sect. 2.6.

2.5 Parameter Configuration

In this section, we investigate how to tune the used parameters in the protocol to maximize the time efficiency which is defined as the ratio of the labeled tag population size to the execution time in each round. The reason for optimizing time efficiency lies in that the higher time efficiency means that the more tags will be labeled per unit time.

The execution time of the current round, defined as T, comprises the time to transmit the CIV and group data. Denote by z the frame size in the labeling phase, for f slots are executed in the screening phase, T can be calculated as

$$T = \lceil \log_2(l+1) \rceil \cdot f \cdot t_0 + z \cdot t_g, \tag{2.1}$$

where t_0 and t_g denote the time for the reader to transmit one bit and group data, respectively.

Let m be the number of tags labeled in the considered round, then the time efficiency in this round, denoted by u, is

$$u = \frac{m}{T} = \frac{m}{\lceil \log_2(l+1) \rceil \cdot f \cdot t_0 + z \cdot t_g}. \tag{2.2}$$

Given (2.2) on u, we next need to find such a pair of f and l that u achieves the maximum. Note that we use u and $u(f, l)$ interchangeably in the rest of the chapter. As m and z and their relationship in the protocol cannot be formulated, it is necessary to search the optimal parameter pair of f and l. For this purpose, we propose a dynamic searching algorithm.

Before introducing the searching algorithm, we first establish an upper bound for f and l, denoted by $\overline{f}$ and $\overline{l}$ respectively, in the following lemma.

Lemma 2 *For* $\forall f > \overline{f}$ *and/or* $l > \overline{l}$, *it holds that* $\hat{u}(f, l) < \hat{u}(\overline{f}, l)$ *and* $\hat{u}(f, l) < \hat{u}(f, \overline{l})$ *where* $\hat{u}(f, l) = \frac{N'}{\lceil \log_2(l+1) \rceil \cdot f \cdot t_0 + G' \cdot t_g}$.

Proof The proof is provided in Appendix 2.C. □

Having derived the upper-bounds of f and l, we get the searching region $[1, \overline{f}] \times [1, \overline{l}]$. To speed up the searching process, we propose a dynamic searching algorithm updating the value of $\overline{f}$ and $\overline{l}$ for the $(Q+1)$-th search from the observations of the Q leading searches. Let f_q, l_q with $1 \leq q \leq Q$ denote each pair of f and l in the first Q searches, we can find the optimal pair (f_{q*}, l_{q*}) contributing to the greatest u in the first Q searches. Given f and l, executing any of AA, c-search-**I** and c-search-**II** will return u. With observations above, we update $\overline{f}$ and $\overline{l}$ by solving the following equations:

$$\text{Update } \overline{f} : u(f_{q*}, l_{q*}) = \hat{u}(\overline{f}, l_{q*}), \qquad (2.3)$$

$$\text{Update } \overline{l} : u(f_{q*}, l_{q*}) = \hat{u}(f_{q*}, \overline{l}). \qquad (2.4)$$

Formally, 2.4 the searching process is illustrated in Algorithm 6. With the input of the number N' of the unlabeled tags, the number G' of groups with unlabeled tags as well as the step size for f, t_l and t_g, Algorithm 6 will output the optimal pair (f^*,l^*) and the maximum time efficiency u^*.

Considering the memory of commercial tags ranges from 32 bits to 8192 bits [12], one cannot use an arbitrary number of seeds, so we denote by l_{act} the maximum seeds a tag can store in its memory. Consequently, we need to update $\overline{l}$ in Algorithm 6 by choosing the minimum one between l_{act} and the solution of (2.4). Note that we set l_{act} to 10 in the simulation.

Moreover, we investigate how the frame size f influences the time efficiency u via the experiment where $l_{act} = 10$ and $N = 10^3$ tags are evenly partitioned into $G = 4, 8, 10$ groups. Specifically, we snapshot the first round of GLMS with c-search-**I** and c-search-**II**. Figures 2.8a, b show that the time efficiency u can be regarded as convex approximately with respective to f. It is thus feasible to employ the gradient method to speed up the search for the optimum f^*.

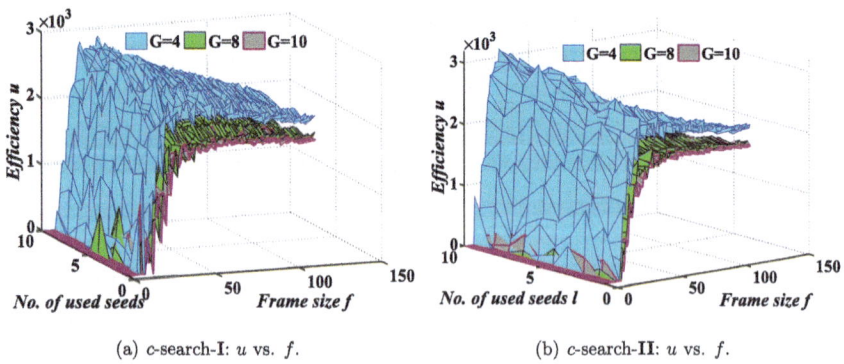

(a) c-search-**I**: u vs. f. (b) c-search-**II**: u vs. f.

Fig. 2.8 c-search-**I** and c-search-**II**: u versus f

2.6 Performance Evaluation

Discussion on Multi-reader case. In large-scale RFID systems deployed in a large area, multiple readers are required to ensure the full coverage for a larger number of tags. To work with multiple readers, we leverage the same approach as [7, 17, 18] that the back-end server synchronizes and schedules all readers such that a multi-reader RFID system operates as the single-reader one. Specifically, the back-end server calculates all the parameters and constructs the CIV involved in the group labeling protocol, and sends them to all readers such that the readers broadcast the same parameters and CIV to the tags.

Explanation on NP-hardness. When $l_{act} = 1$ or the optimum $l^* = 1$, our protocol is degraded to the single-seed protocol which does not need to assign seeds and is not NP-hard. The NP-hard seed assignment problem arises from the employment of multiple seeds. Albeit NP-hardness brings new challenges, we design a series of algorithms running in polynomial time to approximate the optimum and confirm their performance theoretically and experimentally. Moreover, the computation is done in the back-end server which is usually of a high computational capacity.

Potential implementation. Considering the implementation of the proposed protocol, programmable tags, such as those based on WISP hardware, and a USRP-based Software-Defined RFID reader are needed. In order to achieve hashing functionality, hash values are pre-stored in each tag, which is supported by WISP 4, WISP 5, and MSP430. In the scheme implementation, two commands need to be added: 1) TRANSIV that is used to transmit the CIV; 2) QUERPAR that contains the parameters used in the protocol and starts the slot.

Specifically, the reader first sends a TRANSIV commend to broadcast the CIV and then sends a QUERPAR commend. Consider an arbitrary slot j. When a tag receives this command, it starts computing the number by selecting the $\lfloor \log f \rfloor$-bit string starting from the i-th bit in the pre-stored hash value like in [9], where i is the seed value of the j-th position in the CIV. If the number equals to the current slot number, then the tag waits and receives the data sent from the reader.

2.6 Performance Evaluation

2.6.1 Simulation Settings

We evaluate the performance of proposed approaches in comparison with the state-of-the-art solution CCG [8]. We conduct the experiments under both symmetric and asymmetric scenarios with various numbers of groups and group sizes. By symmetric/asymmetric, we mean that the tag population size in each group is identical/different. We use the communication parameters specified in the EPC-global C1G2 standard [15]. Specifically, the data rate from the reader to tags is 26.7 kbps, meaning it takes 37.45 μs for the reader to transmit one bit, so we have $t_1 = 37.45$ μs. We take group ID of $\lceil \log_2 G \rceil$ bits as group data, so we have $t_g = 37.45 * \lceil \log_2 G \rceil$. Besides, we consider the time interval of 302 μs between any two consecutive communications between the reader and tags in the computation of the execution time.

Due to the complexity of AA, we will focus on evaluating the GLMS running the simplified algorithms, namely GLMS with c-search-I and GLMS with c-search-II, but we can measure the performance of AA from Fig. 2.6a in the RFID system of 1000 tags. As discussed in Sects. 2.4.2 and 2.5, we set $c = 40$ for c-search-I, and set $c = 1$ and $l_{act} = 10$ for c-search-II. Albeit using $l_{act} = 10$, we also evaluate its impact on the performance. For simplification, we will use c-search-I and c-search-II in the figures below to stand for GLMS with c-search-I and c-search-II, respectively.

2.6.2 Simulation Results

The performance metric is the communication cost in terms of execution time. We first show the influence of l_{act} with a diverse number of groups G and tags N in the system, and simulate symmetric scenarios with G and the group size varied and proceed to its asymmetric counterparts, subsequently.

Performance evaluation under different l_{act}

Here, we conduct experiments to investigate the impact of l_{act} on GLMS with c-search-I and GLMS with c-search-II. To that end, we simulate scenarios with $N = 100, 1000, 2000, 5000$ tags in the system where the tags are evenly partitioned into $G = 4, 8, 10$ groups, respectively. And the value of l_{act} are set to 10, 15, 20. The simulation results are listed in Table 2.3.

As shown in Table 2.3, the increase in the value of l_{act} reduces the execution time under all settings. Specifically, the performance difference between $l_{act} = 10$ and $l_{act} = 20$ is bigger than that between $l_{act} = 15$ and $l_{act} = 20$ which is less than 3%. More specifically, we observe from the results that the most significant performance difference is about 11% arising between $l_{act} = 10$ and $l_{act} = 20$ for GLMS with c-search-II when $G = 4$ and $N = 2000$. Considering the constraint on the memory capacity of commercial tags as discussed in Sect. 4.4.2 and the tradeoff between the computational complexity and the execution time, we set $l_{act} = 10$ in the subsequent simulations.

Performance comparison under symmetric scenario

This scenario consists of two cases: one is varying the number of the groups and the other is varying the group size.

Case 1. Here we set the total number of the tags $N = 12000$ and $G = 2 : 2 : 10$ with the identical group size. From the results shown in Fig. 2.9a, we can observe that GLMS with c-search-II and GLMS with c-search-I perform better than CCG, with the performance gain of up to 26.8% and 15.9%, respectively. This is because we employ multiple seeds to reduce the transmission of useless slots and c-search-II can furthest exploit the heterogeneous slots

2.6 Performance Evaluation

Table 2.3 Execution time under diverse N, G, l_{act}: studying the impact of l_{act}

Protocol		Vary the number of groups G and l_{act}: (G, l_{act})								
		(4,10)	(4,15)	(4,20)	(8,10)	(8,15)	(8,20)	(10,10)	(10,15)	(10,20)
c-search-I	$N = 100$	0.025	0.025	0.025	0.039	0.037	0.036	0.042	0.041	0.039
	$N = 1000$	0.333	0.311	0.31	0.444	0.426	0.424	0.473	0.456	0.449
	$N = 2000$	0.680	0.636	0.623	0.935	0.877	0.871	0.982	0.928	0.917
	$N = 5000$	1.700	1.647	1.622	2.260	2.240	2.204	2.385	2.287	2.282
c-search-II	$N = 100$	0.024	0.024	0.024	0.037	0.035	0.035	0.041	0.038	0.037
	$N = 1000$	0.32	0.285	0.282	0.431	0.407	0.391	0.455	0.425	0.421
	$N = 2000$	0.629	0.572	0.562	0.890	0.813	0.801	0.946	0.880	0.860
	$N = 5000$	1.581	1.459	1.433	2.209	2.001	1.978	2.331	2.127	2.204

that will become useful. Besides, increasing the number of groups renders more execution time, as more groups reduce the useful slot probability.

Case 2. Here we set $G = 3, 6$ while varying the group size from 500 to 2000, and show the results in Figs. 2.9b, c, respectively. As shown in the pictures, GLMS with c-search-I and GLMS with c-search-II can still finish the group labeling task within the less time than CCG. Especially, with c-search-II, GLMS can save time, under all group size settings, at least 22.5% when $G = 3$, and at least 14.8% when $G = 6$.

Performance comparison under asymmetric scenario

This scenario consists of three cases: the first two cases are the asymmetric counterparts of the symmetric scenarios, i.e., varying the number of the groups and the group size, respectively, and we increase the asymmetry in the third case.

Case 1. In this case, we choose each group size randomly from [100, 2000] while varying G from 2 to 10, and depict the results in Fig. 2.10a. It can be drawn from Fig. 2.10a that c-search-II achieves the best time efficiency and c-search-I performs better than CCG, which results from the ability of our approaches of exploiting more useful slots. Specifically, c-search-II and c-search-I reduce the time up to 34.2% and 24.3%, respectively, in comparison with CCG.

Case 2. In this case, we set the number of the groups to $G = 3, 6$, and choose the group size randomly from $[a, 5000]$ with $a = 125, 625, 1250, 2500$. Figures 2.10b, c depict the simulation results, from which we observe that c-search-II performs best and c-search-I is also better than CCG. Specifically, c-search-II and c-search-I reduce the time cost up to

Fig. 2.9 Performance comparison in the symmetric scenario with the various number of groups and group sizes: smaller execution time means better performance

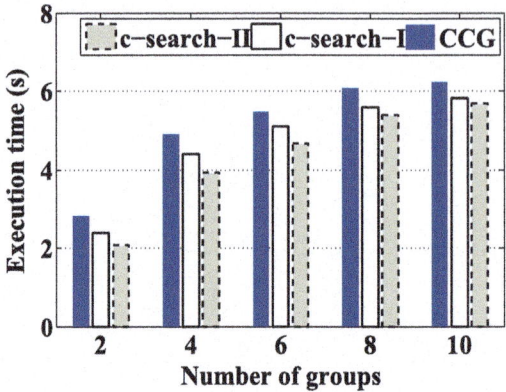

(a) Case 1: $N = 12000$ and various groups G

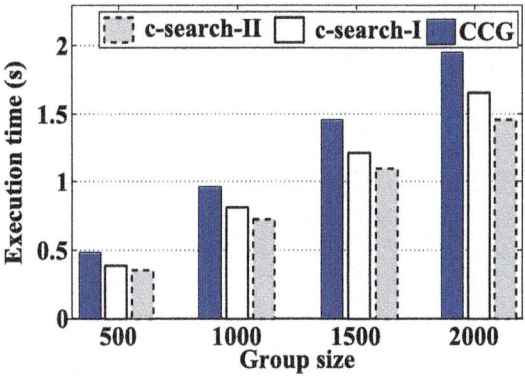

(b) Case 2: $G = 3$, varying group size N_g

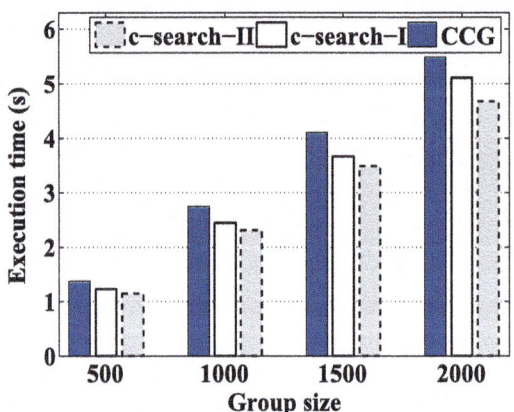

(c) Case 2: $G = 6$, varying group size N_g

2.6 Performance Evaluation

Fig. 2.10 Performance comparison in an asymmetric scenario with the various number of groups and group sizes: smaller execution time means better performance

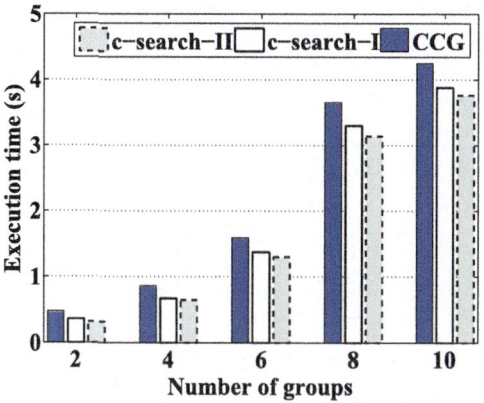

(a) Case 1: various groups G

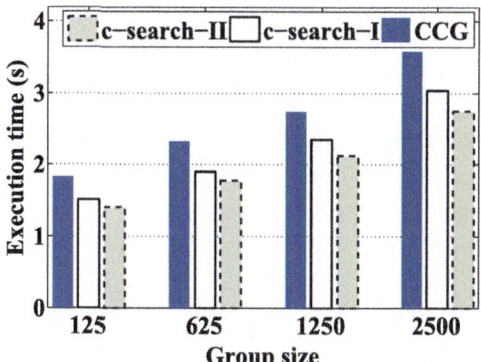

(b) Case 2: $G = 3$, varying group size N_g

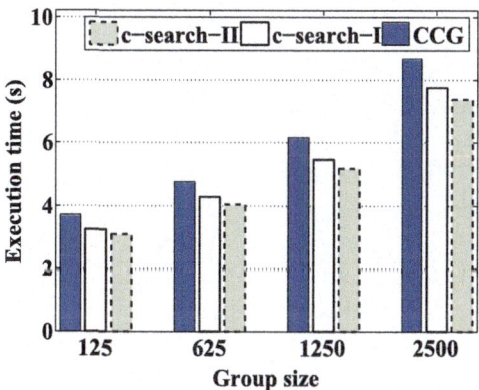

(c) Case 2: $G = 6$, varying group size N_g

Table 2.4 Performance evaluation in case 3

Protocol	Subcase 1	Subcase 2	Subcase 3	Subcase 4
CCG	1.1971	2.9932	1.429	3.1914
c-search-**I**	0.9567	2.5905	1.1704	2.7825
c-search-**II**	0.8719	2.3832	1.0496	2.6052

23.5% and 18.3% when $G = 3$, and up to 17.2% and 12.1% when $G = 6$, respectively, in comparison with CCG.

Case 3. In this case, we also set $G = 3, 6$, but we synthesize the following four subcases by choosing the group size from different ranges: Subcase 1: $G = 3$, we choose the group size randomly for the first group from [100, 500], and from [2000, 3000] for the others. Subcase 2: $G = 6$, we choose the group size randomly for the three groups from [100, 500], and from [2000, 3000] for the others. Subcase 3: $G = 3$, we choose group size randomly for the three groups from [100, 500], [1000, 2000], and [2000, 3000], respectively. Subcase 4: $G = 6$, we choose the group size randomly for the first two groups from [100, 500], from [2000, 3000] for the last two groups, and [1000, 2000] for the others, respectively. As shown in Table 2.4, c-search-**II** and c-search-**I** always outperform CCG. Specifically, CCG spends up to 27.6% and 20.1% time more than ours, respectively, for the transmission of useless slots.

Performance comparison under asymmetric scenario with other distributions

Normal distribution: We consider three cases, each of which has the same number of the groups but has the different group sizes. Specifically, we set $G = 2 : 2 : 10$ in all cases, and each group size follows the normal distribution $N(1000, \delta^2)$ with the standard deviation δ varied from 200 in Case 1 to 400 in Case 2 and to 800 in Case 3. As shown in Fig. 2.11, GLMS with c-search-**II** is the fastest with the less complexity than c-search-**I**, and saves time of up to 27%, 23%, 28% in the three cases, respectively, compared with CCG. Zipfian distribution: Each group size is sampled from [1, 1000] following the Zipfian distribution $Z(1000, 1, G)$ with the number of groups G set to {10, 20, 50, 100}. The performance gain of c-search-**II** over CCG is 31%, 27%, 20%, and 8%, respectively.

2.7 Related Work

Group labeling is a common functionality for many RFID applications. This section presents the prior works on group labeling and the existing multi-seed/hash RFID protocols.

The feasible solutions to the group labeling problem. One straightforward solution is to use the basic polling protocol (BP) [14] where each tag is polled with its group data by the reader one by one. And BIC [19] can label each tag with its group data by informing each

2.7 Related Work

Fig. 2.11 Performance comparison in an asymmetric scenario with the normally distributed group size: smaller execution time means better performance

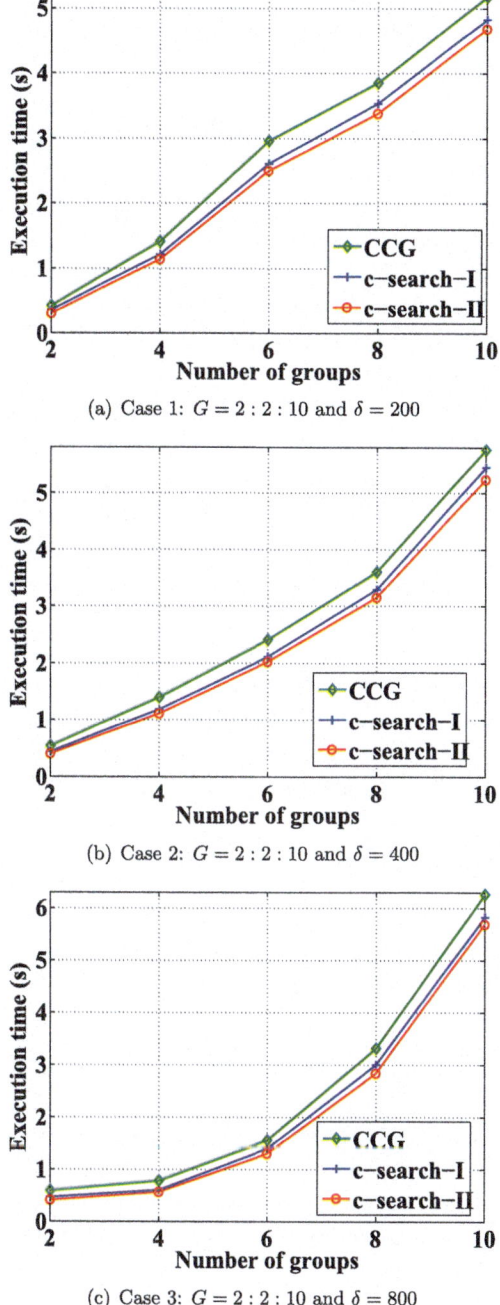

(a) Case 1: $G = 2 : 2 : 10$ and $\delta = 200$

(b) Case 2: $G = 2 : 2 : 10$ and $\delta = 400$

(c) Case 3: $G = 2 : 2 : 10$ and $\delta = 800$

tag of the singleton slot when the tag should wait for its group data. These methods only employ singleton slots such that only one tag can be labeled per slot, as a result, they spend too much time either sending many tag IDs or group data and are thus time-consuming.

To improve time efficiency, the authors in [8] devise three protocols, namely EPG, FIG, and CCG. In EPG, the reader first polls all tags in the same group and sends the group data once. EPG is better than BP for less transmission of group data, however, it still wastes time sending many tag IDs. In FIG, the reader builds a Bloom filter for each group from its tags to filter out the tags of the other groups. Although outperforming EPG, FIG suffers from the false positives of the Bloom filter and has to deactivate the wrong tags by polling, which increases the time cost. To address this problem, CCG allows the reader to send different group data to tags of multiple groups in one round. The reader sends a single indicator vector to inform tags of each slot state such that only the tags in the useful slots will receive their respective group data. Instead of using one seed in CCG, this chapter employs multiple seeds to build a composite indicator vector to further improve the time efficiency.

Multi-seed/hash-based protocols in RFID systems. The multi-seed/hash methods are used to address the information collection and tag monitoring tasks in RFID systems. Chen et al. [9] employs multiple hashes to enable fast information collection. Then, the multi-seed/hash method is used in monitoring the missing tag event and unknown tag event. Specifically, Luo et al. [10] introduce the multi-seed method to detect missing tags in an RFID system. The works [7, 20, 21] address the missing tag detection and identification with multiple hashes. Recently, Gong et al. [11] combined the Bloom filter with the multi-seed method in order to detect the unknown tags fast and reliably. The main novelty of our work is exploiting collision slots instead of only singleton or empty slots in these works. Moreover, we address a different group labeling problem, making the theoretical analysis completely new. We would like to emphasize that this chapter is the first work proving the NP-hardness of SAP arising from the application of multiple seeds and designing the approximation algorithms, which makes our work more challenging.

2.8 Conclusion

This chapter investigates the methods for achieving efficient group labeling. To this end, we proposed a novel multi-seed group labeling protocol GLMS. We found an NP-hard seed assignment problem that arises from the use of multiple seeds. To address this problem, we first introduced an approximation algorithm with a proven competitive ratio and subsequently designed two simplified algorithms that offer lower complexity while maintaining comparable performance. The simulation results demonstrate the superiority of the proposed approaches.

2.A Proof of NP-Hardness

Proof To show the NP-hardness of Problem 1, we prove it reducible from the Maximum coverage problem that is a classic NP-hard problem in polynomial time. Before the formal proof, we first introduce the Maximum coverage problem.

Problem 2 (Maximum Coverage Problem) Consider a set $\mathbb{U}$ of n elements, and a collection $\mathbb{S} = \{\mathbb{S}_1, \mathbb{S}_2, \cdots, \mathbb{S}_h\}$ of h subsets of $\mathbb{U}$ such that $\cup_r \mathbb{S}_r = \mathbb{U}$ where $r = \{1, 2, \cdots, h\}$. Given an integer $k \leq h$, the Maximum coverage problem seeks k subsets from $\mathbb{S}$ maximizing the cardinality of their union.

Next, we show the polynomial reduction by three steps. Given h subsets $\mathbb{S}_r$ of $\mathbb{U}$ and integer $k \leq h$ with which we need to solve Problem 2, we instantiate Problem 1 as below.
Step 1. We first replicate each set $\mathbb{S}_r$ k times and obtain $h \times k$ sets as shown in Fig. 2.12 and $1 \leq r \leq h$.
Step 2. Since in Problem 1, a slot in the CIV should only be assigned one seed, that is, just one set should be selected in each column in Fig. 2.12. To this end, we introduce k dummy sets $\mathbb{U}_1, \mathbb{U}_2, \cdots, \mathbb{U}_k$ such that $\cap_{r'} \mathbb{U}_{r'} = \emptyset$ and $\mathbb{U}_{r'} \cap \mathbb{S}_r = \emptyset$ and $|\mathbb{U}_{r'}| \gg \max_r |\mathbb{S}_r|$ where $1 \leq r' \leq k$ and $|\cdot|$ represents the cardinality of a set. Moreover, we assume that all dummy sets have the same cardinality for simpleness.

Subsequently, let $\mathbb{U}_{r'}$ unite each set in column r' as shown in Fig. 2.13. As $\cap_{r'} \mathbb{U}_{r'} = \emptyset$ and $\mathbb{U}_{r'} \cap \mathbb{S}_r = \emptyset$ and $|\mathbb{U}_{r'}| \gg \max_r |\mathbb{S}_r|$, if a set in column r' is picked, then only choosing a set from another column r'_1 will contribute to more new elements and can thus lead to a greater $|(\mathbb{S}_r \cup \mathbb{U}_{r'}) \cup (\mathbb{S}_{r_1} \cup \mathbb{U}_{r'_1})|$ where $r' \neq r'_1$ and $1 \leq r_1 \leq h$.
Step 3. Recall the objective in Problem 1 that we would like to designate seeds for z useful slots to maximize $u = \frac{|\cup_j C_j|}{t_g(a+z)}$ where $a = \lceil \log_2(l+1) \rceil \cdot f \cdot t_0/t_g$ is a constant. To reduce Problem 2 to Problem 1, we proceed in this step to prove that u reaches its maximum only if k sets are chosen in Problem 1.

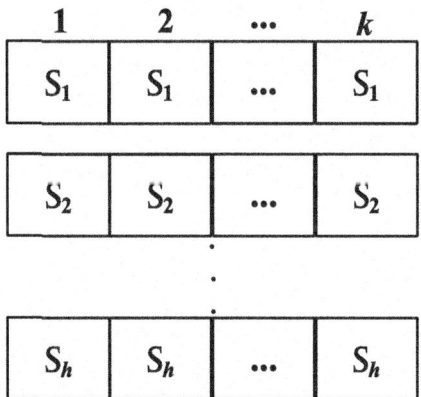

Fig. 2.12 Instantiation of Problem 1

Fig. 2.13 Construction of dummy sets

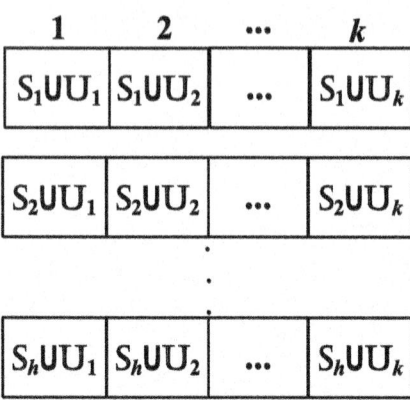

To that end, suppose u reaches its maximum in the case that $k' < k$ sets ($\mathbb{S}_r \cup \mathbb{U}_{r'}$) in Fig. 2.13 have been selected and the selected k' subsets $\mathbb{S}_r$ covers X elements, the time efficiency u is thus computed as:

$$u = \frac{k'|\mathbb{U}_{r'}| + X}{t_g(a + k')}.$$

Adding one more set to the supposed case, we have

$$u' = \frac{(k'+1)|\mathbb{U}_{r'}| + X + Y}{t_g(a + k' + 1)},$$

where Y means the number of new elements contributed by the newly selected $\mathbb{S}_r$. If we can prove that $u' > u$, then the optimum of u achieves if and only if k sets are selected. Let $u < u'$, algebraically, we derive the following condition:

$$\frac{X}{|\mathbb{U}_{r'}|} < a + \frac{(a+k')Y}{|\mathbb{U}_{r'}|}.$$

Since $|\mathbb{U}_{r'}|$ is supposed to be large enough, it is easy to check that the condition in the equation above is established.

After three steps, Problem 1 is polynomially reduced from Problem 2 which is NP-hard, Problem 1 is thus NP-hard. □

2.B Proof of Lemma 1

Proof Let u_{opt} denote the optimal time efficiency of Problem 1 and define m_{opt} and z_{opt} as the optimum for u_{opt}. That is, in the optimal case, there are m_{opt} tags covered by z_{opt} useful slots. For simpleness, we define a function $\Lambda(m, z) = m$ as the number of the tags covered by z slots and use $\phi(z)$ to stand for $t_g(a + z)$, thus u can be expressed as $u = \frac{\Lambda(m,z)}{\phi(z)}$.

Consider that z_{opt} useful slots are selected, we have

$$u_{opt} = \frac{\Lambda(m_{opt}, z_{opt})}{\phi(z_{opt})} \leq \left(\frac{1}{1-\frac{1}{e}}\right) \frac{\Lambda(m, z_{opt})}{\phi(z_{opt})}$$

$$\leq \left(\frac{1}{1-\frac{1}{e}}\right) u_{max},$$

where the first inequality is established by the fact that the number of the tags covered by the greedy approximation algorithm is at least $(1-\frac{1}{e})$ of the optimal solution to the Maximum coverage problem, and the second inequality holds since u_{max} is the maximum time efficiency for any z.

With algebraic operation, Lemma 1 is thus proven. □

2.C Proof of Lemma 2

Proof We employ $\hat{u}(f, l)$ to define the function for time efficiency u with respect to f and l in the case that all N' unlabeled tags from G' groups in the considered round can be labeled within G' slots. Given either f or l, it is easy to check from $\hat{u}(f, l)$ that we cannot enhance $\hat{u}(f, \bar{l})$ or $\hat{u}(\bar{f}, l)$ by increasing l or f. Therefore, the lemma is proven. □

References

1. D. Wu, M. J. Hussain, S. Li, L. Lu, R^2: Over-the-air reprogramming on computational rfids, in *IEEE RFID* (2016), pp. 1–8
2. D. Wu, L. Lu, M.J. Hussain, S. Li, M. Li, F. Zhang, R^3: Reliable over-the-air reprogramming on computational rfids. ACM Trans. Embedded Comput. Syst. **17**(1), 9 (2017)
3. M. Chen, J. Liu, S. Chen, Q. Xiao, Efficient anonymous category-level joint tag estimation, in *IEEE ICNP* (2016), pp. 1–10
4. X. Liu, K. Li, A. X. Liu, S. Guo, M. Shahzad, A. L. Wang, J. Wu, Multi-category RFID estimation, *IEEE/ACM TON* (2016)
5. X. Liu et al., Top-k queries for multi-category RFID systems, in *IEEE Infocom* (2016), pp. 1–9
6. B. Sheng, C.C. Tan, Q. Li, W. Mao, Finding popular categories for RFID tags, in *ACM MobiHoc* (2008), pp. 159–168
7. J. Yu, L. Chen, R. Zhang, K. Wang, On missing tag detection in multiple-group multiple-region rfid systems. IEEE TMC **16**(5), 1371–1381 (2017)
8. J. Liu, B. Xiao, S. Chen, F. Zhu, L. Chen, Fast RFID grouping protocols, in *IEEE Infocom* (2015), pp. 1948–1956
9. S. Chen, M. Zhang, B. Xiao, Efficient information collection protocols for sensor-augmented RFID networks, in *IEEE Infocom* (2011), pp. 3101–3109
10. W. Luo, S. Chen, T. Li, Y. Qiao, Probabilistic missing-tag detection and energy-time tradeoff in large-scale RFID systems, in *ACM MobiHoc* (2012), pp. 95–104

11. W. Gong, J. Liu, Z. Yang, Fast and reliable unknown tag detection in large-scale RFID systems, in *ACM MobiHoc* (2016), pp. 141–150
12. IMPINJ, RFID tag chips. Available: http://www.impinj.com/products/
13. Y. Zheng, M. Li, P-mti: Physical-layer missing tag identification via compressive sensing. IEEE/ACM TON **23**(4), 1356–1366 (2015)
14. Y. Qiao, S. Chen, T. Li, S. Chen, Energy-efficient polling protocols in rfid systems, in *ACM MobiHoc* (2011), p. 25
15. EPCglobal Inc., Radio-frequency identity protocols class-1 generation-2 UHF RFID protocol for communications at 860 mhz–960 mhz version 1.0.9, 2005. Available: http://www.gs1.org
16. V.V. Vazirani, *Approximation algorithms* (Springer-Verlag, Berlin Heidelberg, 2003)
17. M. Kodialam, T. Nandagopal, W. C. Lau, Anonymous tracking using RFID tags, in *IEEE Infocom* (2007), pp. 1217–1225
18. M. Shahzad, A. X. Liu, Expecting the unexpected: Fast and reliable detection of missing RFID tags in the wild, in *IEEE Infocom* (2015), pp. 1939–1947
19. H. Yue et al, A time-efficient information collection protocol for large-scale rfid systems, in *IEEE Infocom* (2012), pp. 2158–2166
20. X. Liu et al., A multiple hashing approach to complete identification of missing RFID tags. IEEE TCOM **62**(3), 1046–1057 (2014)
21. J. Yu, L. Chen, R. Zhang, K. Wang, Finding needles in a haystack: Missing tag detection in large rfid systems. IEEE TCOM **65**(5), 2036–2047 (2017)

Secure Anonymous Group-Wise Writing Scheme for RFID Systems

3

This chapter is devoted to providing anonymous group writing. Our solution involves constructing an approximately random sequence as noise by overlapping transmission data from different tag groups, thereby hiding the original information with low computational complexity. In this context, we propose an Overlapped Bloom Filter (OBF), a compact filter that guarantees time efficiency while improving the security of the group writing. To ensure tags can verify the correctness of decoded group data, the enhanced version, OBF+, introduces the complementary code-based check mechanism to eliminate faulty data. We prototype the system with USRP and programmable WISP tags and conduct extensive simulations to evaluate our approaches in terms of time efficiency, accuracy, and data anonymity.

Chapter roadmap: The rest of this chapter is organized as follows. Section 3.1 explains the motivation for studying the efficiently anonymous group writing and summarizes the contributions. Section 3.2 reviews prior works on group writing and existing approaches to achieving anonymity in RFID transmission. In Sect. 3.3, the system model, including the problem formulation of the anonymous group writing and the motivation from overlapping group data is presented. We then detail the proposed OBF and the advanced OBF+ in Sects. 3.4 and 3.5, respectively. Section 3.6 describes the implementation of the proposed OBF+ for the anonymous group writing using USRP software-defined radio and programmable WISP tags. In Sect. 3.7, we evaluate the performance of proposed approaches in different scenarios. Finally, we conclude this chapter in Sect. 3.8.

3.1 Introduction

One of the fundamental functions in RFID-based systems is group writing, which is to inform each group of the tags of their group data (e.g., group ID). Yet, sending plaintext group data in the prior works discloses the privacy of the systems such as the group ID and password, introducing the risk of being attacked. In this context, one would like to conduct anonymous group writing, which can correctly inform each tag of its group data while guaranteeing the anonymity of the group data against an eavesdropper.

One intuitive solution is to introduce mature encryption algorithms [1] into the existing group writing [2–4] via encrypting the group data. However, the direct encryption has two disadvantages: On the one hand, it needs to embed an integrated encryption/decryption protocol into the original protocol increasing the communication overhead. On the other hand, the tags have to be equipped with the corresponding decryption module increasing the computational complexity, which is unsuitable for the tags of limited energy. Therefore, it is called for a lightweight anonymous group writing protocol to protect the privacy of group data in a time-efficient way.

In this chapter, we propose an Overlapped Bloom Filter-based (OBF) protocol and its enhanced version (OBF+) to achieve efficient anonymous group writing. The two protocols hide the group data to be sent with simple logical operators, e.g., OR and AND. The OBF encodes the data of each tag group with bit overlapping allowed (logical 'OR') at the positions where each tag is mapped, forming an approximately random sequence as a noise at the reader side, and then allows each tag to recover the group data from the received bit sequence via logical 'AND'. On top of the OBF, the OBF+ brings the ability to check the recovered group data to address the fault data by using the complement of the group data. Although expanding the data with its complement increases the frame size reducing the time efficiency of the transmission, this eliminates the incorrectly recovered group data while improving the anonymity of the group data, enhancing the reliability of the anonymous group writing. Our contribution can be summarized as follows.

- We formulate the largely unaddressed anonymous group writing problem in RFID systems and provide solutions from the perspective of initiatively adding harass via bit overlapping into the transmission bit sequence.
- We present two concrete protocols namely OBF and the enhanced version OBF+ to construct and recover the 'noise' sequence. Both of them can achieve the required accuracy of the recovered data under anonymous transmission, and the OBF+ achieves higher reliability and stronger anonymity.
- We optimize the protocol performance by the optimum parameters derivation. The analytical results reveal the relationship among the time efficiency, the accuracy of the recovered group data, and the transmission anonymity.
- We prototype the system with the software-defined radio reader and programmable WISP tags. The experimental results confirm the feasibility of our protocol in practice.

- We conduct extensive simulations to evaluate the performance of the proposed protocols. The results show that the time cost of the OBF+ is about twice as the OBF, but there is no fault data and the crack probability decreases by two orders of magnitude when there are 5000 tags.

3.2 Related Work

Group writing is an important functionality for managing multi-task backscatter systems with different types of RFID tags. In this section, we present the prior works about group writing and the existing anonymity of RFID transmission.

Polling protocol [5] can access the targeted tags thus the reader directly transmits the group data to the tags one by one. The BIC [2] can transmit the group data to each tag at the singleton slot mapped by the tag. The above methods address the problem of group data transmission, but they only just achieve one tag per slot in the transmission instead of a batch of the tags, leading to low time efficiency and low privacy. To improve the time efficiency, Liu et al. [3] employ the slots mapped by the tags from the same group, thus not only the singleton slots but also some collision slots can be used for group data transmission. Yu et al. [6] pick multiple seeds to build a composite indicator vector to further reduce the useless slots in the transmission. Liu et al. [7] design a writing bundling scheme to bundle multiple writes up and execute them together in a burst mode, which greatly reduces the number of selects and thereby amplifies the write throughput. Liu et al. [8] prove an algorithm-independent bound of time efficiency in terms of the minimum number of needed 'Select' commands and propose a writing scheme based on the designed pseudo-ID of tags. None of these works can achieve the anonymity of the data transmission.

To protect the data transmission between the readers and the tags, Gao et al. [1] use the Rabin public key cryptography algorithm, which verifies the signature process via square multiplication and modulo operations to prevent eavesdropping by unauthorized users. Yang et al. [9] introduce the chaotic sequences to encrypt/decrypt transmission data so that the transmitted data after the encryption is similar to the white noise sequence improving the anonymity of the data transmission. Each tag has its unique chaotic key stream sequence to encrypt the data. Regrettably, it only just achieves one tag per slot in the transmission reducing the time efficiency. Conducting these protocols requires more computation resources beyond the capability of tags. The main novelty of our work is to introduce interference via overlapping the data of different groups to improve the anonymity of each group without adding new special modules like encryption/decryption as the conventional anonymity methods in RFID systems. Moreover, we address a different group writing problem, making the theoretical analysis new.

3.3 System Model and Problem Formulation

3.3.1 System Model

We consider an RFID system consisting of one reader[1] and a large number of tags. The reader is connected with a back-end server which has a powerful computational capability. For the purpose of simplification, we treat the reader and the server as an entity and just call it the reader. Moreover, each tag has a unique ID and performs computations such as hashing function. All tags' IDs in the system are recorded by the reader.

The communication between the reader and tags follows the rule of 'Listen-before-talk'. At the beginning of the communication, the reader broadcasts the commands containing the parameters such as the seed to the tags, and each tag conducts specified operations according to the received commands. In the data transmission, for example, the reader broadcasts the commands and the parameters including the frame size f and a seed s at first and constructs the data frame such as the bloom filter **BF** to transmit. Then, each tag uses its ID and the received seed s to generate one pseudo-random value via the hash function as $h = (H(\mathbf{ID}, s) \mod f)$. The random value from the hash function shows the start point where to read in the bloom filter. Finally, the tags execute the next step according to the received commands (i.e., compare, respond, or wait for the next commands). Note that the hash function used in our study is ideal and its output obeys uniform distribution and is nonreversible.

The downlink (i.e., reader-to-tags) transmission is continuous [14]. For simplicity, we denote T_d as the time duration of a one-bit broadcasting slot. Consider an arbitrate broadcasting slot, it conveys either '1' or '0'. We assume that the unauthorized users do not have any prior knowledge of the tag's ID and the transmission data but they can perfectly synchronize with the reader's transmission thus eavesdropping the data broadcasted from the reader.

3.3.2 Problem Formulation

We are interested in anonymous group writing in an RFID system. In this chapter, n tags belong to g groups and each group has its corresponding group data, i.e., $\mathbf{GD}_j$ for $1 \leq j \leq g$, and the length of the group data is l-bit.

The problem of the anonymous group writing over multiple tag categories is to make an arbitrary tag success in recovering l-bit group data from the transmission sequence under the accuracy requirement and an anonymity requirement within the minimum time. The accuracy of the group writing in the system is defined as

---

[1] For multiple readers, we treat them as a single virtual reader as in [10–13]. Specifically, the back-end server configures the parameters and sends them to all readers. Consequently, the back-end server can synchronize the readers and we can logically consider them as a whole.

3.3 System Model and Problem Formulation

$$\mathbb{E}\left[\frac{n_{cor}}{n}\right] \geq \alpha, \tag{3.1}$$

where n_{cor} is the number of the tags correctly recovering group data and α is the required accuracy. The execution time of the protocol is related to the frame size f, i.e., the length of the transmitting bit sequence, so we should minimize the frame size under the required accuracy, i.e., $f_{opt}(\alpha)$.

For the anonymity of the group data, on the one hand, the constructed sequence for the transmission should approach a binomial distribution sequence where each bit in the sequence is nearly irrelevant to prevent from cracking, i.e., the probability of a bit being '1' or '0' in the transmission sequence should be $\frac{1}{2}$. We denote by P_{un} the anonymity which measures the probability of incorrect crack from the transmission sequence. Intuitively, P_{un} is positively related to the frame size and the group data length, that said

$$P_{un} \propto \frac{1}{f_{opt}(\alpha)}\left(\frac{1}{2}\right)^l. \tag{3.2}$$

In this context, we would like to maximize P_{un}. Therefore, there exists a trade-off between the time efficiency and the anonymity. Table 3.1 summarizes the used main notations.

3.3.3 Overview of Our Solutions

In the scene of the anonymous group writing, the reader should not only support the simultaneous data transmission for the multiple groups of the tags but also prevent unauthorized users from cracking the group data from the transmitted data.

At the reader side, we directly hash the group data in bits and record them at the positions of a bloom filter where each tag is mapped by ORing with the bits that were recorded in these positions. As a result, the different group data will overlap with each other, achieving simultaneous writing for multiple groups while hiding the original information. At the tag side, they conduct an 'AND' operation to recover the original tag of each group from the received overlapped sequence. We refer to this new bloom filter as the Overlapped Bloom Filter.

Moreover, motivated by the feature of the complement in binary sequence and the bitwise 'OR' and 'AND', we could construct a special sequence that consists of a number and its complement number to be a judgment factor to remove the fault data resulting from incorrectly recovered group data due to the false positives of the bloom filter and the overlapping of the data. The extended length of the data also increases the anonymity of the data.

Table 3.1 Main notations

Symbols	Description
n	The number of the tags in our system
g	The number of the groups in our system
f	The frame size of hashing
s	The seed used in hash function
k	The number of hash mappings conducted by each tag
$h_{i,m}$	The m-th hash value of the i-th tag
$\mathbf{BF}[a]$	The content of the a-th element in the bloom filter $\mathbf{BF}$
l	The bit length of each group data
$\mathbf{GD}_j$	The group ID of the j-th group
f_o	The frame size of the hashing in the OBF
$P_{wrg}^{(j)}$	The probability of writing the fault group data in the j-th group in the OBF
P_{cor}	The expected rate of recovering the correct group data in OBF
f_{eo}	The frame size of the hashing in the OBF+
q	The q-th round of multi-round writing in the OBF+
k_q	The number of hash mappings for a tag in the q-th round with the OBF+
P_{ce}	The probability of recovering the correct group data for each round in the OBF+
T_d	The period of broadcasting in a bit

3.4 OBF: Overlapped Bloom Filter-Based Group Writing

In this section, we first introduce the OBF and analyze how to tune the parameters for performance optimization.

3.4.1 Motivation

Inspired by the variant of the bloom filter which consists of an array of cells each containing a fixed number of bits, we achieve group writing by hashing the group data into cells, leading to the simultaneous writing for multiple groups. The major shortage of this variant is its time inefficiency as the minimum unit is a cell (i.e., multiple bits) instead of one bit. Therefore, one intuition is storing group data in bits instead of cells, that said, different group data will overlap with each other by bitwise 'OR' operation, hiding the original information. The

3.4 OBF: Overlapped Bloom Filter-Based Group Writing

advantage also lies in that each tag can conduct an 'AND' operation to recover its group data from the overlapped bit sequence. Now, we verify the feasibility of our idea. Let get start with the rule of Boolean Algebra.

Lemma 1 *A, B and C are the binary sequence with identical length. According to the distributive law, we have*

$$A\&(B|C) = (A\&B)|(A\&C), \tag{3.3}$$

where '&' means the logical bitwise 'AND' operator, and '|' means the logical bitwise 'OR' operator.

Following the above law, we can conduct a conclusion.

Theorem 2 *Denote an l-bit binary sequence by $\mathbf{D}$, and k random l-bit binary sequences by $\mathbf{c}_m$ for $m = 1, 2, \cdots, k$. We construct new sequences $\{\mathbf{e}_1, \mathbf{e}_2, \cdots, \mathbf{e}_k\}$ via making the sequence $\mathbf{D}$ and each random sequence $\mathbf{c}_m$ to do operator $|$, i.e., $\mathbf{e}_1 = \mathbf{D}|\mathbf{c}_1, \mathbf{e}_2 = \mathbf{D}|\mathbf{c}_2, \cdots, \mathbf{e}_k = \mathbf{D}|\mathbf{c}_k$. The result of all these sequences $\mathbf{e}_m$ doing operator & can be written as*

$$\begin{aligned} \mathbf{F} = \mathbf{e}_1 \& \mathbf{e}_2 \& \cdots \& \mathbf{e}_k &= (\mathbf{D}|\mathbf{c}_1) \& (\mathbf{D}|\mathbf{c}_2) \& \cdots \& (\mathbf{D}|\mathbf{c}_k) \\ &= \mathbf{D}|(\mathbf{c}_1 \& \mathbf{c}_2 \& \cdots \& \mathbf{c}_k). \end{aligned} \tag{3.4}$$

Thus, the original sequence $\mathbf{D}$ can be correctly recovered when the $(\mathbf{c}_1 \& \mathbf{c}_2 \& \cdots \& \mathbf{c}_k)$ equals all zero sequence or the sequence whose bit '1' position is the subset of bit '1' positions in the sequence D. For instance, for the sequence $\mathbf{D} = (111001)_2$, if the $(\mathbf{c}_1 \& \mathbf{c}_2 \& \cdots \& \mathbf{c}_k)$ equals each case of $(110000)_2$, $(001000)_2$, $(111001)_2$, the sequence $\mathbf{D}$ can be correctly recovered as $\mathbf{D} = (111001)_2$.

Proof For the first two sequences $\mathbf{e}_1$ and $\mathbf{e}_2$, we have

$$\mathbf{F}_1 = \mathbf{e}_1 \& \mathbf{e}_2 = (\mathbf{D}|\mathbf{c}_1) \& (\mathbf{D}|\mathbf{c}_2) = \mathbf{D}|(\mathbf{c}_1 \& \mathbf{c}_2). \tag{3.5}$$

Then, ANDing $\mathbf{F}_1$ with $\mathbf{e}_3$, yields $\mathbf{F}_2 = \mathbf{F}_1 \& \mathbf{e}_3$. Similarly, after k 'AND' operations, we can eventually get

$$\mathbf{F} = \mathbf{F}_{k-1} = \mathbf{F}_{k-2} \& \mathbf{e}_k = \mathbf{D}|(\mathbf{c}_1 \& \mathbf{c}_2 \& \cdots \& \mathbf{c}_k). \tag{3.6}$$

Here, we prove the feasibility of hiding the sequence $\mathbf{D}$ via logical operator 'OR' and recovering the sequence $\mathbf{D}$ via logical operator 'AND'. □

Since the mapping in the bloom filter by the hash function is random, the inserted sequence might randomly overlap with each other, which is regarded as the target sequence $\mathbf{D}$ overlapped by a random sequence $\mathbf{c}_m$. Therefore, the overlapped data is recoverable for each tag. So far, we proved the effectiveness of our idea by inserting group data into the bloom filter.

Obviously, the tags can decode the correct group data if $(\mathbf{c}_1 \& \mathbf{c}_2 \& \cdots \& \mathbf{c}_k)$ is a sequence of full zeros or the positions of the bit '1' in the $(\mathbf{c}_1 \& \mathbf{c}_2 \& \cdots \& \mathbf{c}_k)$ is the subset of that in the **D**. Note that the $\mathbf{c}_m$ approaches a random sequence and the $(\mathbf{c}_1 \& \mathbf{c}_2 \& \cdots \& \mathbf{c}_k)$ is near a full-zero sequence with the increasing of the k. Therefore, the key to improving writing accuracy lies in increasing the probability that $(\mathbf{c}_1 \& \mathbf{c}_2 \& \cdots \& \mathbf{c}_k)$ is a zero sequence. The configuration of the parameters satisfying the required accuracy of the group writing will be discussed in Sect. 3.4.3.

3.4.2 Protocol Description

Our proposed Overlapped Bloom Filter can be described as follows. We start at the view of the reader.

First, the reader updates grouping rules through the users' requirement and then groups n tags into their corresponding groups and allows each group a group data such as Group ID (c.f. $\mathbf{GD}_j$ for $1 \leq j \leq g$).

Second, the reader constructs an f_o-bit bloom filter **BF** that all bits are initialized to '0'. For the i-th tag of the j-th group in the system where $1 \leq i \leq n$, each hashing value from the hash functions with the seed s, ID, and the frame size f_o is $h_{i,m}$ for $1 \leq m \leq k$. The $h_{i,m}$ points out the starting position to insert in the bloom filter for this tag, and we update the content of the bloom filter via $\mathbf{BF}\left[h_{i,m}: \mod (h_{i,m}+l-1, f_o)\right] = \mathbf{BF}\left[h_{i,m}: \mod (h_{i,m}+l-1, f_o)\right] | \mathbf{GD}_j$. After k insertions, we complete the encoding of the i-th tag's group data into the bloom filter via the operator 'OR'.

Third, once the overlapped bloom filter is constructed from the mapping of all tags, the reader broadcasts the seed s, the frame size f_o, and the overlapped bloom filter to tags.

At the tag side, each tag receives the parameters from the reader and uses them to calculate its k hashing values under the seed s, ID and the frame size f_o. The tag then extracts each l-bit binary sequence starting from its mapped positions from the received overlapped bloom filter. The tag obtains a sequence by combining these k l-bit sequences via the bitwise 'AND'. This recovered sequence is regarded as the group data for this tag.

Note that the accuracy of group writing is not 100% after the above steps since the OBF is a probabilistic algorithm. The reader can predict the tags that received the incorrect group data. To inform the left tags of their group data, the reader has to conduct multi-round writing. Specifically, the reader informs each tag of the wrong data with the 'select' command and conducts a new round of writing with OBF. The time cost is acceptable since the number of the incorrectly received tags is small with the required writing accuracy.

One of the challenges in the overlapped bloom filter is how to tune the frame size f_o and the number of the mapping positions k to optimize the time efficiency and anonymity given the required rate of recovering the correct group data. We will address this in the Sect. 3.4.3.

3.4 OBF: Overlapped Bloom Filter-Based Group Writing

3.4.3 Parameters Optimization

We here introduce how to set the parameters used in the overlapped bloom filter so that the reliability of recovering group data can be satisfied while the overall execution time is minimal and then analyze the anonymity of the transmission.

In our system, there exists n tags and g groups in our system. In the j-th group, it involves n_j tags, and its corresponding l-bit group data is $\mathbf{GD}_j$. We assume w_j as the number of '1' in the $\mathbf{GD}_j$. The length of the bloom filter $\mathbf{BF}$ is f_o and all bits in this filter are initialized to '0'. The time cost of transmitting the seed value and the frame size can be negligible compared with the time cost of transmitting the bloom filter. Therefore, the total execution time is determined by the overlapped bloom filter transmission duration, which is expressed as

$$T_{whole_1} = f_o T_d. \tag{3.7}$$

Obviously, the key to minimizing the total execution time is to find the minimum frame size under the constrain of $P_{cor} \geq \alpha$, where P_{cor} is the expected rate of recovering the correct group data in our system and it is expressed as

$$P_{cor} = 1 - \mathbb{E}\left[P_{wrg}^{(j)}\right] \geq \alpha, \tag{3.8}$$

where $P_{wrg}^{(j)}$ is the probability of incorrectly decoding the group data. Now, the goal is to find the minimum frame size f_o to satisfy $1 - \mathbb{E}\left[P_{wrg}^{(j)}\right] \geq \alpha$. Let's start with the first group. The probability of an arbitrary bit in $\mathbf{BF}$ mapped by a tag in the first group is $\frac{1}{f_o}$. Considering we will insert l-bit sequence into $\mathbf{BF}$ via the operator 'OR', the probability of an arbitrary bit in $\mathbf{BF}$ covered by a tag in the first group is $\frac{l}{f_o}$. Furthermore, the probability of an arbitrary bit holding '0' after once mapping is $1 - \frac{l}{f_o} \cdot \frac{w_1}{l} = 1 - \frac{w_1}{f_o}$. Hence, the probability of an arbitrary bit in $\mathbf{BF}$ still being '0' after mapping of all tags of the first group is

$$P_{grp_1} = \left(1 - \frac{w_1}{f_o}\right)^{kn_1}. \tag{3.9}$$

After the entire tag set has been inserted their group data into the overlapped bloom filter, the probability of an arbitrary bit in $\mathbf{BF}$ maintaining '0' is

$$P_0 = \prod_{j=1}^{g} P_{grp_j} = \prod_{j=1}^{g} \left(1 - \frac{w_j}{f_o}\right)^{kn_j}. \tag{3.10}$$

Thus, the probability that an arbitrary bit is '1' after bitwise 'AND' of the k l-bit binary sequences extracted from the overlapped bloom filter can be expressed as

$$P_1 = (1 - P_0)^k = \left(1 - \prod_{j=1}^{g}\left(1 - \frac{w_j}{f_o}\right)^{kn_j}\right)^k. \tag{3.11}$$

Decoding fails when some '0' bits change to '1' after the bitwise operation 'OR' and 'AND'. Hence, the probability of the fault data for the j-th group can be expressed as

$$P_{wrg}^{(j)} = 1 - (1 - P_1)^{l-w_j}, \tag{3.12}$$

where $l - w_j$ represents the number of the zero bits in $\mathbf{GD}_j$, and $(1 - P_1)^{l-w_j}$ means that these zero bits in $\mathbf{GD}_j$ do not change after bitwise 'OR' and 'AND' operation.

According to the (3.12), $P_{wrg}^{(j)}$ is positively related to P_1 with the fixed w_j. In other words, we will get the minimum value of $P_{wrg}^{(j)}$ when we minimize P_1.

Theorem 3 *Given the number of the tags n_j in each group, the number of bit '1' w_j in each $\mathbf{GD}_j$, the number of groups g, the relationship between the k and the f_o should satisfy*

$$k = -\frac{\ln 2}{\ln b}, \tag{3.13}$$

where $b = \prod_{j=1}^{g}\left(1 - \frac{w_j}{f_o}\right)^{n_j}$.

Proof We donate $b = \prod_{j=1}^{g}\left(1 - \frac{w_j}{f_o}\right)^{n_j}$ and $P_1 = (1 - b^k)^k$. To minimize the P_1, we derive the partial differential function of $\ln P_1$ with respect to the k. This partial differential function can be expressed as

$$\frac{\partial \ln P_1}{\partial k} = \frac{\partial}{\partial k} k \ln\left(1 - b^k\right)$$

$$= \ln\left(1 - b^k\right) - \frac{b^k \ln b^k}{1 - b^k}. \tag{3.14}$$

We will get the minimum values of P_1 when the $\frac{\partial \ln P_1}{\partial k} = 0$ under $b^k = \frac{1}{2}$. Thus, the relationship between k and f_o satisfies $k = -\frac{\ln 2}{\ln b}$ where $b = \prod_{j=1}^{g}\left(1 - \frac{w_j}{f_o}\right)^{n_j}$. □

Recall (3.8), we substitute (3.12) into it and rewrite as

$$P_{cor} = 1 - \mathbb{E}\left[P_{wrg}^{(j)}\right] = (1 - P_1)^{l - \mathbb{E}[w_j]} \geq \alpha, \tag{3.15}$$

and $\mathbb{E}\left[w_j\right]$ can be expressed as

$$\mathbb{E}\left[w_j\right] = \frac{1}{n}\sum_{j=1}^{g} n_j w_j. \tag{3.16}$$

3.4 OBF: Overlapped Bloom Filter-Based Group Writing

Substituting (3.11) and (3.13) into (3.15) thus yields

$$\sum_{j=1}^{g} n_j \ln\left(1 - \frac{w_j}{f_o}\right) \geq \frac{(\ln 2)^2}{\ln\left(1 - \alpha^{\frac{1}{l - \mathbb{E}[w_j]}}\right)}. \tag{3.17}$$

The key is to find the minimum value of f_o making the above inequality hold. Therefore, we obtain the minimum frame size f_o for the group data transmission, which minimizes the total execution time of the protocol.

Now, we analyze the anonymity of the overlapped bloom filter-based protocol. On the one hand, $P_0 = \frac{1}{2}$ under the optimum configuration means that the overlapped bloom filter approaches a binomial distribution sequence that each bit is independent, increasing the concealment of the transmission. On the other hand, the unauthorized users can obtain the frame size of the bloom filter, the seed, and the data size since the reader broadcasts these parameters with plaintext. But, the critical of the problem is to stop the group data from being cracked. For the l-bit sequence, the number of the possible sequences is 2^l. The expected ratio of the number of '0' and '1' in the sequence is 1, hence the expected number of the '0' and the '1' in any sequence is $l/2$. For a tag's any l-bit sequence which would be inserted into the overlapped bloom filter, the probability of this inserted sequence remaining unchanged after overlapping, i.e., the probability of this inserted sequence being cracked, is

$$P_{un1} = P_{a0}^l + P_{a0}^{\frac{l}{2}}(1 - P_{a0})^{\frac{l}{2}}, \tag{3.18}$$

where P_{a0}^l means the inserted position of the tag's group data in the overlapped bloom filter is l-bit zero sequence after the other sequences inserted and $P_{a0}^{\frac{l}{2}}(1 - P_{a0})^{\frac{l}{2}}$ means that the inserted position is the identical sequences as this tag's group data after the other tags inserted. P_{a0}^l can be expressed as

$$P_{a0} = \frac{P_0}{\left(1 - \frac{\mathbb{E}[w_j]}{f_o}\right)} = \frac{1}{2\left(1 - \frac{\mathbb{E}[w_j]}{f_o}\right)}. \tag{3.19}$$

Recall (3.13), we have $P_{a0} \approx P_0 = 1/2$ since f_o is much larger than $\mathbb{E}[w_j]$. Therefore, the anonymity of the overlapped bloom filter-based protocol, i.e., the probability of the unauthorized users being unable to crack the group data from the overlapped bloom filter is

$$P_{ano1} = 1 - \frac{1}{f_o} P_{un1} \approx 1 - \frac{1}{f_o}\left(\frac{1}{2}\right)^{l-1}. \tag{3.20}$$

From the above analysis, the anonymity of the overlapped bloom filter is determined once the frame size f_o is determined given the required rate of correctly recovering the group data.

Let us take an example to interpret the overlapped bloom filter where we aim to transmit group IDs to tags. There exists $n = 3$ tags and $g = 3$ groups. The length of each group ID is $l = 3$ and they are $(110)_2$, $(101)_2$, $(011)_2$ (c.f. Fig. 3.1a). The elements of each group is shown as Fig. 3.1a. Following Theorem 1 to configure the parameters, we have $f_o = 20$ and $k = 3$ and the anonymity of the overlapped bloom filter is 0.9688 since the expected ratio of the number of '0' and '1' in these group IDs is $1/2$. First, the reader constructs the overlapped bloom filter based on inserting all group IDs into the overlapped bloom filter via operator 'OR' (c.f. Fig. 3.1b). Each tag calculates $k = 3$ positions with the received parameters and combines 3 3-bit sequences extracted from the received bloom filter through bitwise operator 'AND' (c.f. Fig. 3.1d). The result is regarded as its group ID by each tag (c.f. Fig. 3.1e).

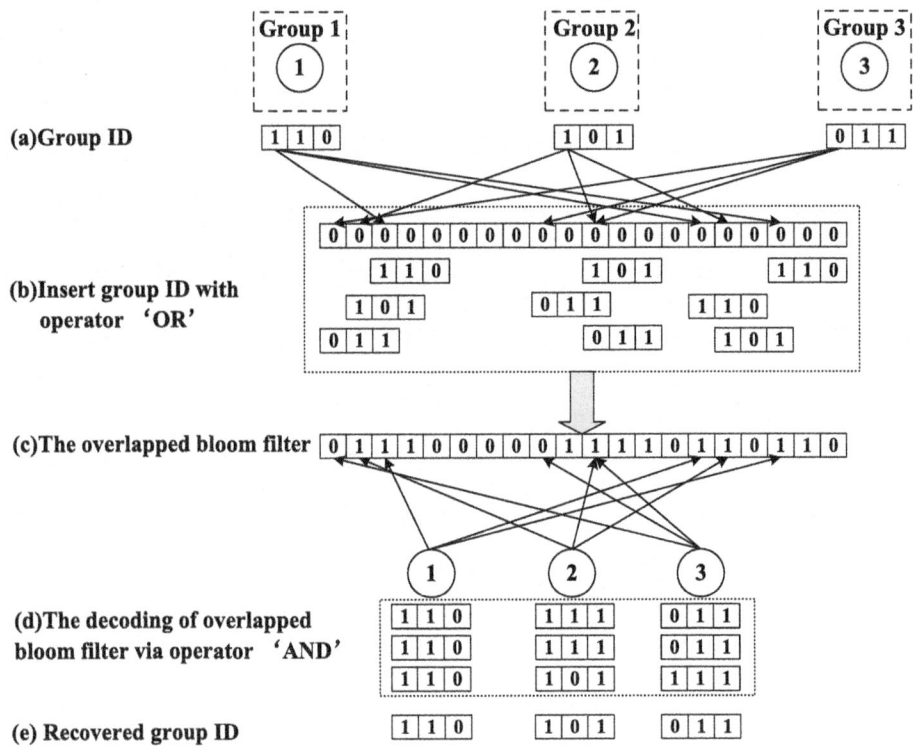

Fig. 3.1 Illustrating the anonymous group writing with OBF

3.5 OBF+: An Enhanced Solution

In this section, we introduce the enhanced version of the OBF to further improve the accuracy of the recovery data and the anonymity of the transmission.

3.5.1 Motivation

As mentioned in (3.4), our overlapped bloom filter suffers from fault recovery data due to that ($c_1 \& c_2 \& \cdots \& c_k$) cannot always be a zero sequence or be the sequence whose 1-bit positions are the subsets of those of **D**. In the OBF, tags cannot actively know whether their recovered group data is correct, so the reader has to spend extra time executing the command 'select' to inform these tags before conducting a new round of writing. To address this limitation, we need a mechanism to enable each tag to verify the correctness of its recovered group data, improving the reliability of the group data transmission. Our solution is to add a check part to the group data. Considering the characteristics of the fault data, we construct a newly inserted sequence by adding the complement of the group data behind itself. If the ratio of the number of the recovered bit '1' and bit '0' in the recovered sequence equals 1, then the decoded group data is correct. Each tag would find the data incorrect once the ratio is not equal to 1.

The advantages of extending the group data with its complement are three-fold. (1) Due to the complement of each group's data, the tags can judge the validity of the recovered sequence, thus enabling multi-round writing operations to improve the accuracy of the writing. (2) The new sequence removes the difference in the accuracy across groups since the number of bit '1' in the new sequence of each group is identical, simplifying the parameter optimization. (3) The probability of the transmission data being cracked decreases because of the increased length of the inserted sequence, improving the transmission anonymity.

Specifically, Let us take an example to illuminate the group data extended with its complement. For a binary sequence $(11010101)_2$, its complement sequence is $(00101010)_2$. Therefore, our extended sequence is $(11010101\ 00101010)_2$. If the recovered sequence after a bitwise 'OR' and 'AND' is $(11010101\ 00101010)_2$, we will treat the former 8-bit $(11010101)_2$ as the group data. If the recovered sequence after a bitwise 'OR' and 'AND' is $(11010111\ 00101010)_2$ where the 7-th bit and the 15-th bits are both '1', the tag could detect fault data and abandon it.

3.5.2 Protocol Description

Our OBF+ can be described as follows. First, the reader groups n tags into g groups and allows each group an l-bit group data (c.f. $\mathbf{GD}_j$ for $1 \leq j \leq g$) according to the requirement of the users.

Second, the reader constructs the enhanced overlapped Bloom filter namely OBF+. The rule of deciding the starting position to insert a binary sequence is the same as the OBF does. The difference is that we insert each group data with its complement followed. Hence, the length of the inserted whole binary sequence is $2 \cdot l$ rather than l in the OBF. The insertion can be expressed as $\mathbf{BF}\left[h_{i,m}: \mod (h_{i,m} + 2l - 1, f_{eo})\right] = \mathbf{BF}\left[h_{i,m}: \mod (h_{i,m} + 2l - 1, f_{eo})\right] | \{\mathbf{GD}_j, \overline{\mathbf{GD}_j}\}$, where f_{eo} is the frame size of the OBF+, and $h_{i,m}$ is the m-th hashing values of i-th tag, and $\overline{\mathbf{GD}_j}$ means the complement sequence of $\mathbf{GD}_j$. Therefore, we insert the i-th tag's group data and its complement into the OBF+ via the operator 'OR'. After the OBF+ has been built, the reader broadcasts the seed s, the frame size f_{eo}, and the OBF+ to the tags. Since the reader has all tags' IDs, the reader can predict the tags that fail to correctly recover the group data by checking whether the former l-bit and the latter l-bit are complementary. For these tags whose group data are unsuccessfully recovered, the reader repeats the above steps to construct another bloom filter from the mapping of these tags with unsuccessful group data.

At the tag side, each tag's group data is initialized to a zero sequence. After receiving the parameters from the reader, each tag first calculates its k hashing values with its ID, the seed s, and the frame size f_{eo}. The tag then extracts $(2l)$-bit binary sequences starting from its corresponding hashing values in the received bloom filter. The tag obtains a recovered sequence after conducting the bitwise 'AND' among these k sequences. It then checks whether the former l-bit and the latter l-bit are complementary. If this holds, the former l-bit is recognized as the group data. Otherwise, the tag knows that the decoded group data is wrong, and would wait for the next writing.

We set a threshold for the maximum writing rounds of the reader since each tag can judge the validity of the recovered group data, enabling the multi-round writing to improve the accuracy. When the writing finishes, the reader would report the IDs of the left tags that fail to recover their group data correctly. The key left here is to configure such parameters as the number of writing rounds and the frame size.

3.5.3 Parameters Optimization

After the multi-round writing, the probability of recovering the correct group data at each round P_{ce} could be smaller than the requirement α, so it is adequate to set the smaller frame size at each round, this however would degrade the anonymity. Therefore, we can proceed in two cases: Optimizing the time efficiency and optimizing the anonymity.

In case 1, we need to find out the minimum frame size of the transmission improving the time efficiency. The time cost of transmitting the seed values and the frame sizes can be negligible compared with that of transmitting the bloom filter. Thus, the overall execution time of the OBF+ is

3.5 OBF+: An Enhanced Solution

$$T_{whole_2} = \sum_{q=1}^{r} f_{eq} T_d, \qquad (3.21)$$

where r is the maximum rounds for bloom filter transmission and f_{eq} is the frame size for the q-th transmission. The key is to configure r and f_{eq} to minimize execution time, and both of them are related to the probability of recovering the correct group data P_{ce} for each round.

Theorem 4 *Given the required accuracy of recovering the correct group data, the maximum rounds for the reader's transmission r and the probability of recovering the correct group data P_{ce} for each round satisfies*

$$r \geq \frac{\ln(1-\alpha)}{\ln(1-P_{ce})}. \qquad (3.22)$$

Proof The number of the tags that unsuccessfully recover their group data after the q-th transmission under the same probability of recovering the correct group data P_{ce} for each round is $n_{eq} = n(1 - P_{ce})^q$. Thus, the number of the left tags after r-round transmission should satisfy $n_{er} \leq n(1-\alpha)$. Therefore, we obtain $r \geq \frac{\ln(1-\alpha)}{\ln(1-P_{ce})}$. □

This theorem shows the relationship between r and P_{ce} under the requirement of accuracy in our protocol. Now the problem is to find the relationship between the P_{ce} and its corresponding frame size f_{eq} for the q-th transmission.

The probability of an arbitrary bit holding '0' after once hashing is $1 - \frac{2l}{f_{eq}} \cdot \frac{l}{2l} = 1 - \frac{l}{f_{eq}}$ since the number of bit '1' and the number of bit '0' are identical in the inserted sequence. Hence, the probability of an arbitrary bit in the q-th bloom filter maintaining '0' after all sequences inserted can be written as

$$Pe_0^{(q)} = \left(1 - \frac{l}{f_{eq}}\right)^{k_q n_{eq}} = \left(1 - \frac{l}{f_{eq}}\right)^{k_q n(1-P_{ce})^{q-1}}.$$

Thus, the probability that an arbitrary bit is '1' after bitwise 'AND' operation of the k_q $2l$-bit binary sequences extracted from the bloom filter can be expressed as

$$Pe_1^{(q)} = (1 - Pe_0^{(q)})^{k_q}$$
$$= \left(1 - \left(1 - \frac{l}{f_{eq}}\right)^{k_q n(1-P_{ce})^{q-1}}\right)^{k_q}. \qquad (3.23)$$

The unsuccessful recovery happens when the former l-bit sequence and the latter l-bit sequence are not complementary in the recovered sequence. That said, there are bits '0' changing to bit '1'. Recall that l bit '0' in the correctly recovered sequence, the probability that none of them has been changed is $\left(1 - Pe_1^{(q)}\right)^l$. Therefore, recovering the correct group data P_{ce} for the q-th round can be expressed as

$$P_{ce} = \left(1 - Pe_1^{(q)}\right)^l$$
$$= \left(1 - \left(1 - \left(1 - \frac{l}{f_{eq}}\right)^{k_q n(1-P_{ce})^{q-1}}\right)^{k_q}\right)^l. \quad (3.24)$$

Due to the same solution described in Sect. 3.4.3, we obtain the relationship between k_q and f_{eq} as

$$k_q = -\frac{\ln 2}{\ln b} \approx \frac{f_{eq} \ln 2}{n(1 - P_{ce})^{q-1} l}, \quad (3.25)$$

where $b = \left(1 - \frac{l}{f_{eq}}\right)^{n(1-P_{ce})^{q-1}}$, and we have $\left(1 - \frac{l}{f_{eq}}\right)^{n(1-P_{ce})^{q-1}} \approx e^{-\frac{nl(1-P_{ce})^{q-1}}{f_{eq}}}$. (3.24) can be rewritten as

$$(1 - P_{ce})^{q-1} \ln\left(1 - (P_{ce})^{\frac{1}{l}}\right)$$
$$= \frac{f_{eq} \ln 2}{nl} \ln\left(1 - \left(1 - \frac{l}{f_{eq}}\right)^{\frac{f_{eq} \ln 2}{l}}\right) \approx -\frac{f_{eq}}{nl}(\ln 2)^2.$$

Then, we have

$$f_{eq} = -\frac{nl}{(\ln 2)^2}(1 - P_{ce})^{q-1} \ln\left(1 - (P_{ce})^{\frac{1}{l}}\right). \quad (3.26)$$

Substituting (3.26) into (3.21) yields

$$T_{whole_2} = \sum_{q=1}^{r} f_{eq} T_d = -\frac{\alpha n l T_d}{P_{ce}(\ln 2)^2} \ln\left(1 - (P_{ce})^{\frac{1}{l}}\right).$$

Obviously, we can obtain the minimum execution time when we have $\frac{dT_{whole_2}}{dP_{ce}} = 0$. Thus, the optimum value of P_{ce} should be satisfy that

$$-\frac{1}{l} P_{ce}^{\frac{1}{l}} = \left(1 - P_{ce}^{\frac{1}{l}}\right) \ln\left(1 - P_{ce}^{\frac{1}{l}}\right). \quad (3.27)$$

Theorem 5 *There is only one non-zero solution to the equation expressed as follows, and it falls into $[1 - e^{a-1}, 1)$.*

$$F(x) = ax + (1 - x) \ln(1 - x) = 0, \quad (3.28)$$

where $0 < a \leq 0.5$ and $0 < x < 1$.

3.5 OBF+: An Enhanced Solution

Proof The first-order and the second-order derivation of Eq. (3.28) can be written as

$$\frac{dF(x)}{dx} = a - \ln(1-x) - 1,$$

$$\frac{d^2F(x)}{dx^2} = \frac{1}{1-x} > 0.$$

We have $\frac{dF(x)}{dx} = 0$ when $x = 1 - e^{a-1}$ and $\frac{dF(x)}{dx} > 0$ when $x > 1 - e^{a-1}$. Substituting this value into (3.28) yields $a - e^{a-1} < 0$. Since we have $F(x) > 0$ when x approaches 1, the non-zero solution falls into $\left[1 - e^{a-1}, 1\right)$. □

Following the theorem, we can obtain the proper value of P_{ce} in the range of $\left[\left(1 - e^{\frac{1}{l}-1}\right)^l, \alpha\right]$ which minimizes the time cost of the transmission.

Next, we analyze the anonymity of the OBF+. Although the size of the sequence expands to $2l$, the number of the possible sequences is also 2^l since the latter l-bit is the complement of the former l-bit. As the same as the OBF, the bloom filter in the OBF+ also approaches a binomial distribution sequence under the above configuration and each bit in the bloom filter is independent. The ratio of the number of the '0' and the '1' in the sequence holds 1, and the number of the '0' and the '1' in any sequence is l. The probability of a bit being '0' in the q-th bloom filter excluding a tag's sequence inserted can be expressed as

$$P_{ea0}^{(q)} = \left(1 - \frac{l}{f_{eq}}\right)^{k_q \left(n(1-P_{ce})^{q-1} - 1\right)} \approx \frac{1}{2}. \qquad (3.29)$$

Thus, the probability of the sequence being cracked is

$$P_{un2}^{(q)} = \left(P_{ea0}^{(q)}\right)^{2l} + \left(P_{ea0}^{(q)}\right)^l \left(1 - P_{ea0}^{(q)}\right)^l = \left(\frac{1}{2}\right)^{2l-1}.$$

Therefore, the anonymity of the enhanced overlapped bloom filter, i.e., the probability of the unauthorized users being unable to crack the group data from the q-th enhanced overlapped bloom filter is

$$P_{ano2}^{(q)} = 1 - \frac{1}{f_{eq}} P_{un2}^{(q)} = 1 - \frac{1}{f_{eq}} \left(\frac{1}{2}\right)^{2l-1}. \qquad (3.30)$$

From the above analysis, the anonymity of the OBF+ is determined by P_{ce} since f_{eq} is determined by P_{ce}. Therefore, we can select the proper value of P_{ce} maximizing the frame size for the anonymity preference. We use the mean value of $P_{ano2}^{(q)}$ measuring the transmission anonymity.

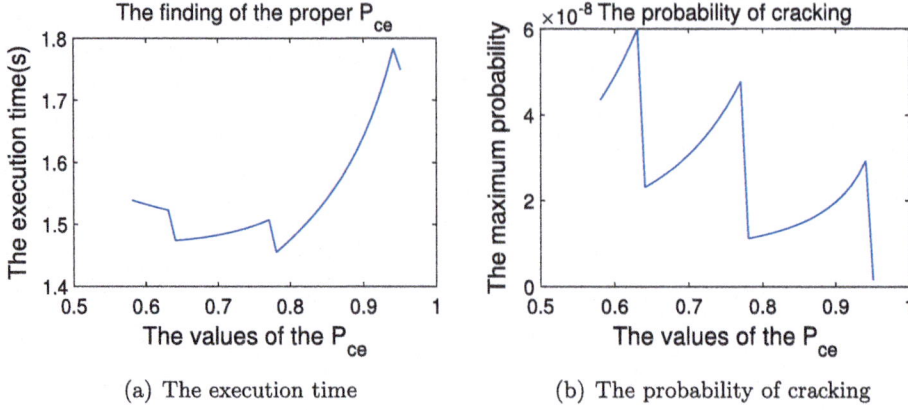

(a) The execution time (b) The probability of cracking

Fig. 3.2 The example of finding proper value of P_{ce} with $\alpha = 95\%$

We will take an example to explain how to select parameters with different preferences in a system consisting of 100 groups and 10 tags in each group (i.e., total 1000 tags in the system), as shown in Fig. 3.2. The downlink transmission rate is 40.97kb/s and the required accuracy of recovering the group data is $\alpha = 0.95$. For each P_{ce}, we obtain its corresponding rounds r and the frame size f_{eq} for the q-th round. For minimizing the time cost, the optimum value of P_{ce} is $P_{ce}^* = 0.78$, the number of wring rounds is $r = 2$, and the frame sizes are $f_{eo1}^* = 48677$ and $f_{eo2}^* = 10855$, respectively. Under this setting, the mean anonymity is $P_{ano2} = 1 - 6.9 \times 10^{-9}$. For maximizing the anonymity, the required probability of recovering the correct group data is $P_{ce}^* = 0.95$, and its corresponding frame size is $f_{eo}^* = 71679$ and the corresponding anonymity is $P_{ano2} = 1 - 1.7 \times 10^{-9}$.

Let us take an example to summarize the process of the OBF+. Consider an RFID system of $g = 3$ groups each having one tag. The reader would transmit the group ID to each tag. The length of each group ID is set to $l = 2$, namely $(01)_2$, $(10)_2$, $(11)_2$ (c.f. Fig. 3.3a). The elements of each group is shown as Fig. 3.3a. Following the parameter configuration rule, we have $r = 1$, $f_{eo} = 20$ and $k = 3$. First, the reader constructs the OBF+ by inserting group IDs and their complements (c.f. Fig. 3.3b) via operator 'OR' (c.f. Fig. 3.3c). Receiving the OBF+, each tag maps itself to $k = 3$ positions, and combines these 3 4-bit sequences through the bitwise operator 'AND' (c.f. Fig. 3.3e). Before acknowledging its group ID, each tag checks whether the former 2-bit and the latter 2-bit are complementary (c.f. Fig. 3.3f). If so, the former 2 bits are the group ID of the tag (c.f. Fig. 3.3g). Otherwise, the tag will regard the recovered group ID $(00)_2$ as wrong.

3.6 Implementation

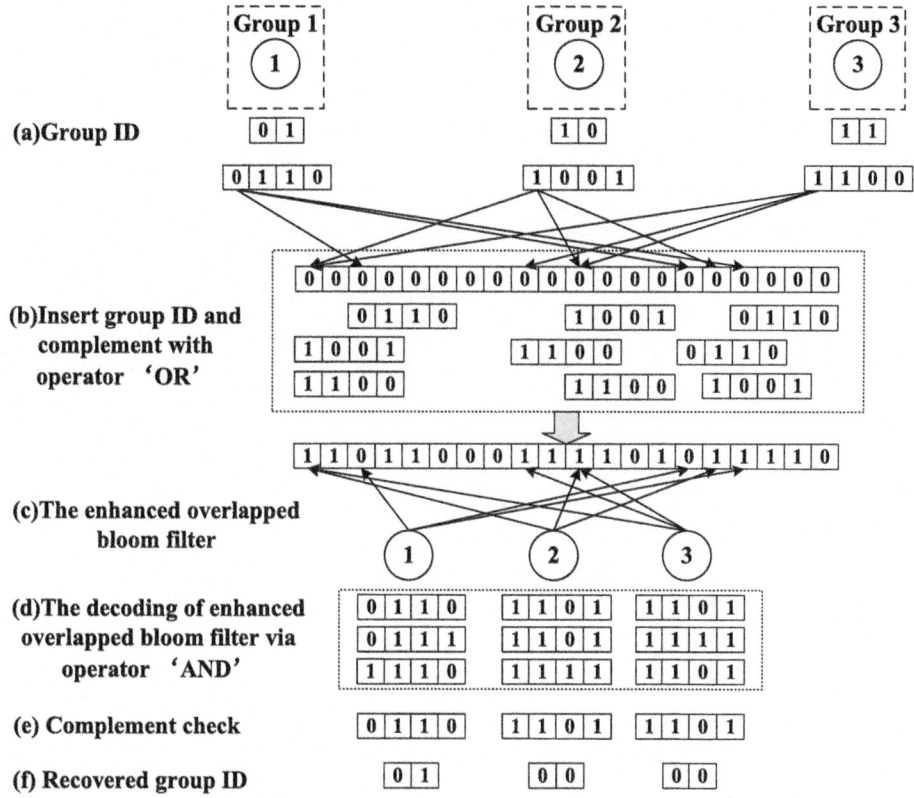

Fig. 3.3 The example of group data transmission based on the OBF+

3.6 Implementation

In this section, we implement our prototype for anonymous group writing using USRP software-defined radio and programmable WISP tags. We select the OBF+ as the protocol used in the prototype since the OBF+ is more reliable thus being more appropriate in practice.

3.6.1 Experimental Setup

As the commodity readers do not transmit a specific binary sequence for the transmission of group data, we customize the Software-Define-Radio(SDR) reader based on NI USRP-2920 software radio to broadcast the bloom filter to tags. The SDR reader uses a USRP WBX daughterboard as the front end, which operates at the center frequency of 915 MHz. Besides, the daughterboard is connected to two Laird S9028PCR circular polarized antennas for receiving and transmitting radio frequency signals respectively. Then We connect the

NI USRP-2920 to a laptop and send the PHY layer symbols to the laptop for the software processing based on the GNURadio. The operating system is Ubuntu 16.04.2 LTS.

Since the tags need to decode the bloom filter from the SDR reader, we implement the tags based on the WISP hardware. The WISP tag is equipped with an ultra-low power MSP430 microcontroller which is able to store the bloom filter and complete the operator 'AND' to decode the bloom filter.

We have an SDR reader and four WISP tags. We divide these tags into two groups each containing two tags. The group data are '01' and '10', respectively. We conduct experiments in our lab and arrange the distance between WISP tags and the SDR reader within 20cm. Moreover, the communication between the WISP tags and the SDR reader still follows the C1G2 standard, we only extend the WISP tags and the SDR reader with the functions of group data transmission. For example, we add new fields into the query command including the number of the hash functions (set to 3), the length of the bloom filter (set to 12), and the constructed bloom filter.

3.6.2 Implementation of the Anonymous Group Writing

The SDR reader broadcasts the modified query command including the bloom filter constructed according to the Sect. 3.5.2. The WISP tags enter the inventory round after receiving the modified query command. For the original fields, the WISP tags follow the C1G2 standard [14]. Then, each WISP tag correctly extracts its own group data if the former part and the latter part of the recovered sequence are complements. Otherwise, the tags will keep silent and wait for the next round. Figure 3.4a plots the PHY layer symbols observed at the SDR reader. Figure 3.4b shows the detail of the bloom filter in the modified query command and Fig. 3.4c shows the details of the four busy slots. Obviously, the ratio of the number of bit '1' and bit '0' approaches $\frac{1}{2}$ and the anonymity is more than 0.99 according to (3.30). As shown in Fig. 3.4c, these four WISP tags successfully recover the group data. Therefore, the OBF+ achieves anonymous group writing in the practical system.

3.7 Performance Evaluation

In this section, we evaluate the performance of the proposed OBF and OBF+ in terms of the accuracy, the execution time, and the anonymity. The timing parameters in the simulation follow the C1G2 standard [14]. Specifically, the communication from the reader to the tags is consecutive transmission. The transmission rate is 40.97kb/s hence a broadcast slot is $T_d = 24.4\mu s$. Each group ID is the binary sequence of its serial number, e.g., the j-th group ID $\mathbf{GD}_j = j$, and the length of the binary sequence is determined by the number of groups g, e.g., $l = \lceil \log_2(g+1) \rceil$. The parameters like the frame size of the bloom filter are set according to the theoretical analysis. In the simulation, we verify the effectiveness of these

3.7 Performance Evaluation

Fig. 3.4 The signal pattern of our prototype

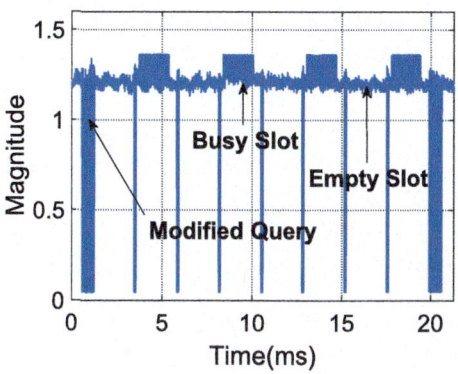

(a) The overview of the signal pattern

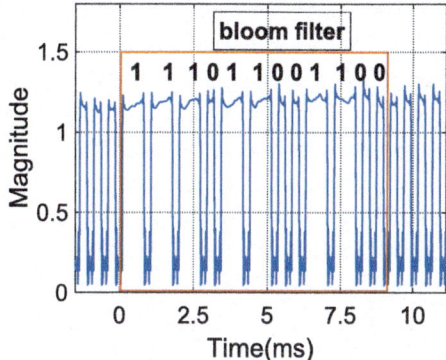

(b) The bloom filter in query command

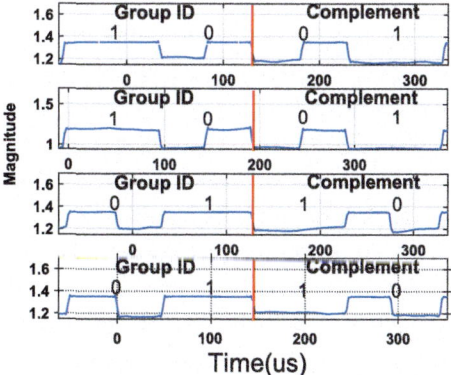

(c) The group data responded from the tags

two protocols in addressing the anonymous group writing problem. The results are obtained from 1,000 independent runs.

Performance Verification: We here verify the time efficiency, the recovery accuracy, the rate of the incorrect group ID, and the anonymity of the proposed protocols under five scenarios. In the simulation, the required recovery accuracy varies from $\alpha = 95\%$ to $\alpha = 99\%$ in the first two scenarios and is fixed to $\alpha = 95\%$ in the latter three scenarios. We use the cracking probability of group data as the metric to gauge the anonymity. The lower the cracking probability means the greater the anonymity. We also show the performance of the OBF+ with different optimization objectives, i.e., time efficiency and anonymity, in the last scenario.

In the first scenario, we study the impacts of the number of overall tags in the system. There are 10 tags in each group, and the number of the total tags varies from 1,000 to 5,000. The simulation results under different required accuracies are depicted in Figs. 3.5 and 3.6. The results show that the proposed OBF and OBF+ can satisfy the required accuracy rate. Yet, these two protocols have to spend more time on group ID transmission as the number of tags increases when there would be more groups. Although the OBF is more time-efficient, the fault data degrades the reliability of the transmission. On the contrary, the rate of recovering incorrect group ID from the OBF+ is always 0 since the OBF+ can detect the error and then abandon the fault data. Moreover, the OBF+ has a lower cracking probability, leading to stronger anonymity of the group data transmission.

In the second scenario, we investigate the impact of the number of tags in each group. To this end, we set the total number of the tags in the system as 1,000 and vary the number of the tags in each group from 1 to 100. Each group size is identical. We can draw from Fig. 3.7 and Fig. 3.8 the similar conclusions as in the first scenario that both OBF and OBF+ can achieve the required accuracy of recovering group ID, but OBF+ is more reliable and more anonymous.

In the third scenario, we further investigate the impact of the random number of the tags in each group under $\alpha = 95\%$. We also set the total number of the tags in the system as 1,000 varying the number of the groups in the system from 5 to 100. Specifically, We randomly classify the tags into g_r groups each consisting of a random number of the tags. We run 100 times and obtain the mean value as the evaluated results with the g_r groups. We can draw a similar conclusion as the second scenario from Fig. 3.9 that both the OBF and the OBF+ can achieve the required accuracy of recovering group ID, but the OBF+ is more reliable and more anonymous.

In the fourth scenario, we verify the OBF and the OBF+ in a large-scale system under $\alpha = 95\%$. Figure 3.10 shows the performance of our two proposed protocols with the number of overall tags in the system given the identical group size (i.e., 10 tags in each group). Moreover, Fig. 3.11 depicts the execution time when the total number of the tags is 30000 and the number of the tags in each group is random. It can be observed that the execution time increases but the cracking probability decreases when the system scale extends.

3.7 Performance Evaluation

Fig. 3.5 The performance of the OBF and the OBF+ with $\alpha = 95\%$ and the number of the overall tags in the system varied

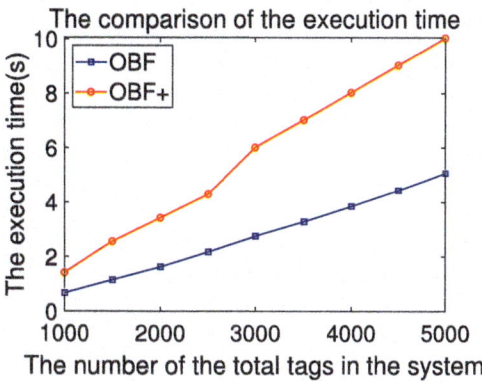

(a) The execution time

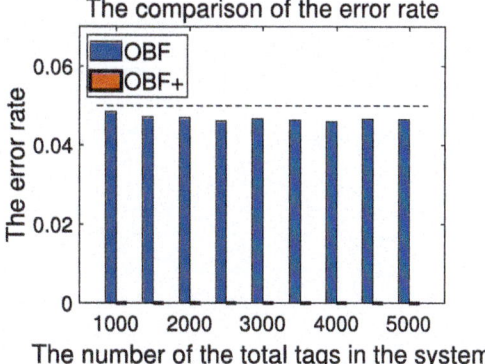

(b) The rate of recovering incorrect group ID

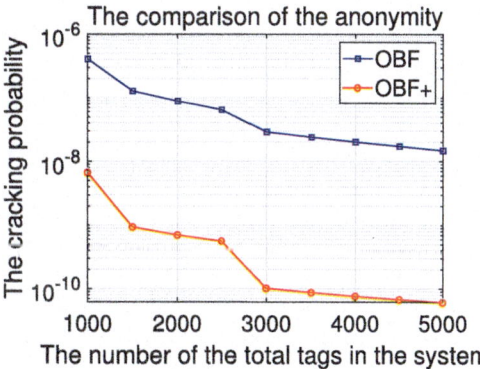

(c) The cracking probability

Fig. 3.6 The performance of the OBF and the OBF+ with $\alpha = 99\%$ and the varying number of the overall tags in the system

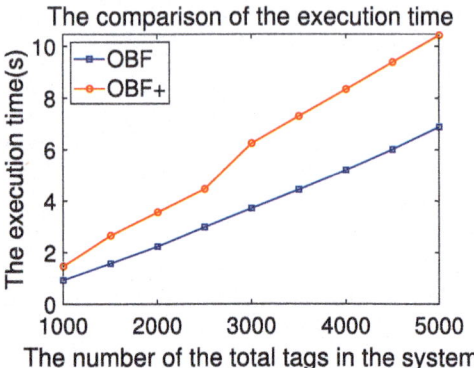

(a) The execution time

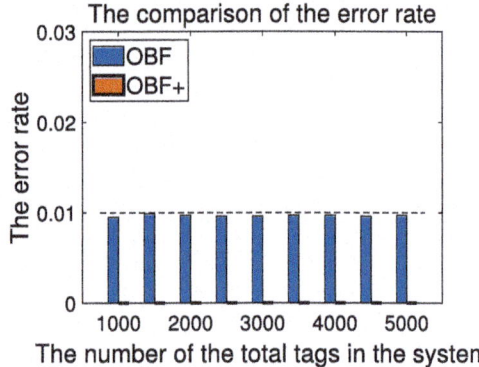

(b) The rate of recovering incorrect group ID

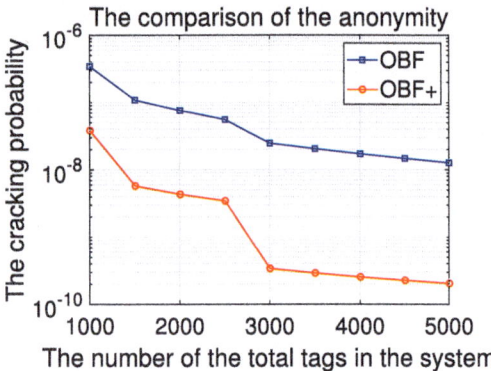

(c) The cracking probability

3.7 Performance Evaluation

Fig. 3.7 The performance of the OBF and the OBF+ with $\alpha = 95\%$ and the number of the tags in each group varied from 1 to 100

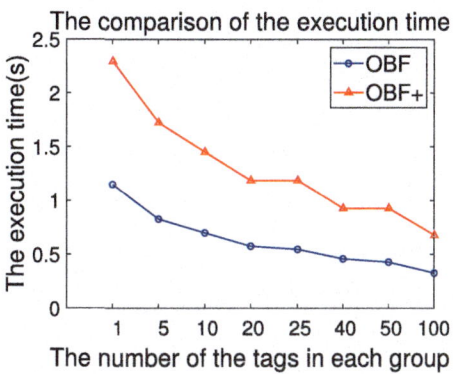

(a) The execution time

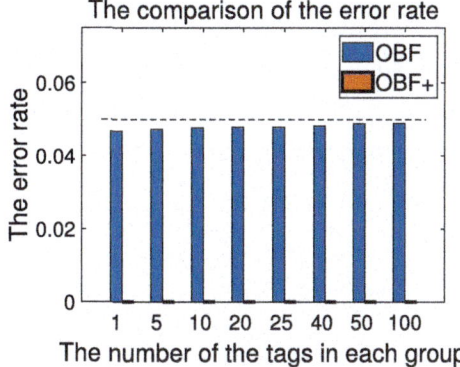

(b) The rate of recovering incorrect group ID

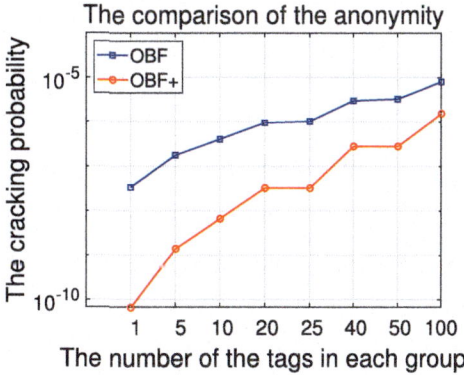

(c) The cracking probability

Fig. 3.8 The performance of the OBF and the OBF+ with $\alpha = 99\%$ and the number of the tags in each group varied from 1 to 100

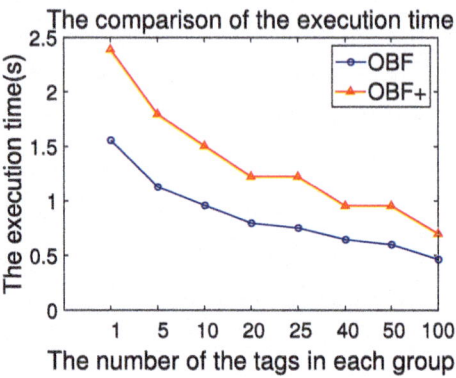

(a) The execution time

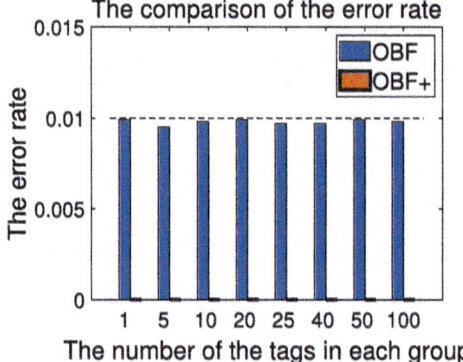

(b) The rate of recovering incorrect group ID

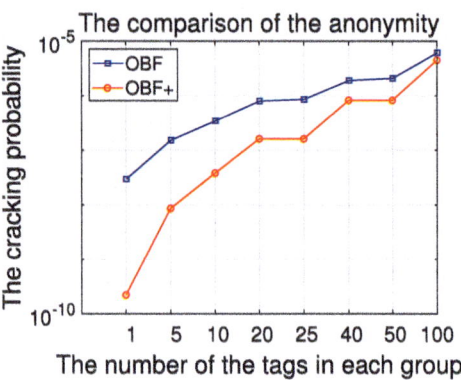

(c) The cracking probability

3.7 Performance Evaluation

Fig. 3.9 The performance of the OBF and the OBF+ with the random number of tags in each group under the accuracy of $\alpha = 95\%$ and the number of total tags is fixed to 1000

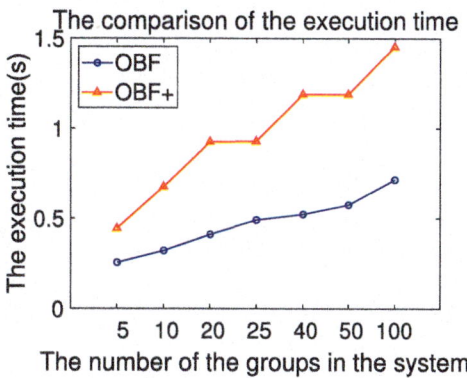

(a) The execution time

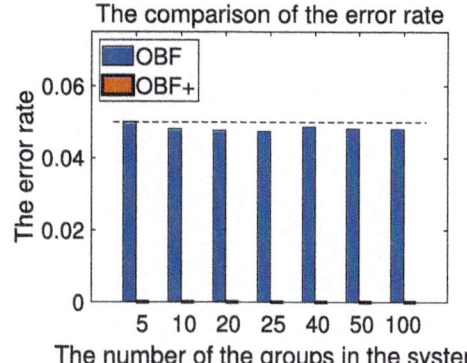

(b) The rate of recovering incorrect group ID

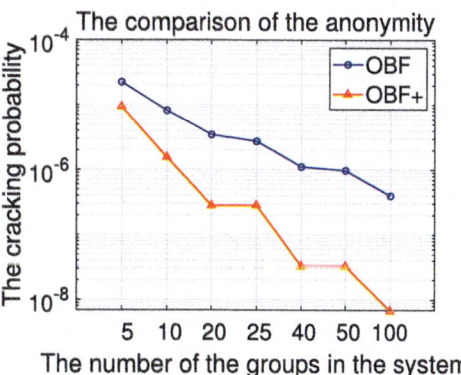

(c) The cracking probability

Fig. 3.10 The performance in the large-scale system where 30000 tags are grouped evenly under $\alpha = 95\%$

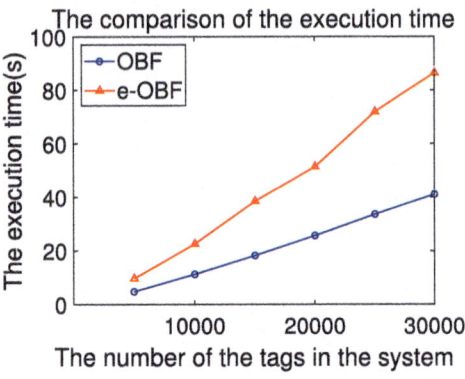

(a) The execution time

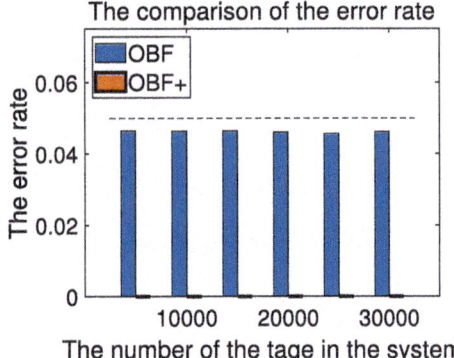

(b) The rate of recovering incorrect group ID

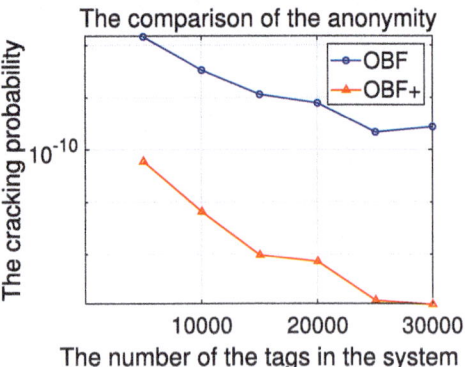

(c) The cracking probability

3.7 Performance Evaluation

Fig. 3.11 The performance on the random number of the tags in each group under $\alpha = 95\%$ and 30000 tags in total

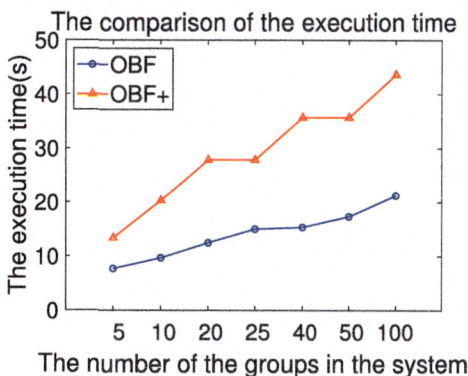

(a) The execution time

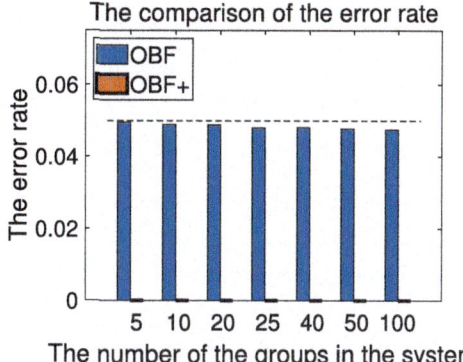

(b) The rate of recovering incorrect group ID

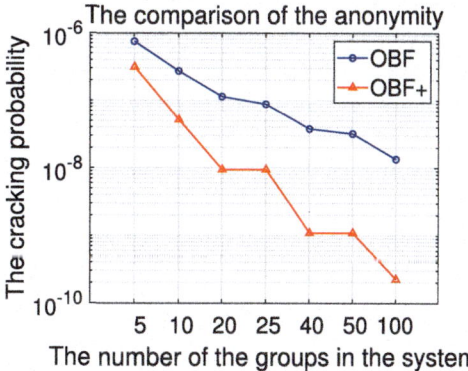

(c) The cracking probability

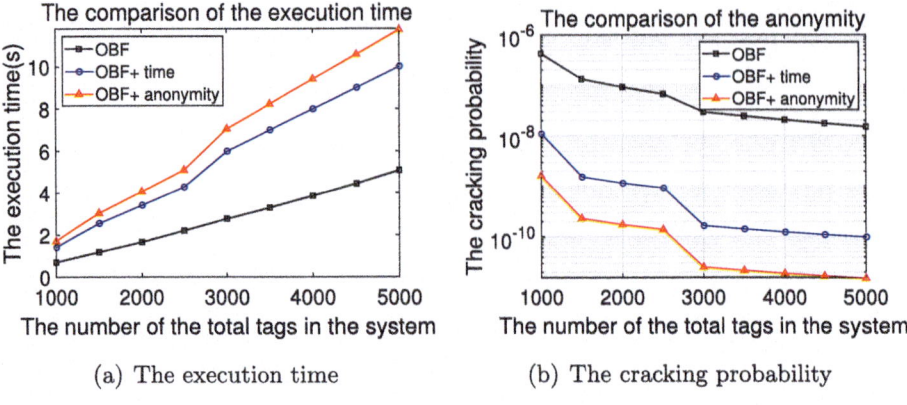

Fig. 3.12 The performance when $\alpha = 95\%$ and each group contains 10 tags

In the fifth scenario, we evaluate the OBF and the OBF+ that work with different optimum strategies (i.e., time cost or anonymity). The rate of recovering the correct group ID is set to $\alpha = 95$. Figure 3.12a illustrates the execution time where the number of tags in each group is identity(i.e., 10 tags in each group). Figure 3.12b depicts the cracking probability. Obviously, the OBF+ improves the anonymity of the protocol at the expense of time efficiency. Therefore, we can make a trade-off between the time cost and anonymity according to the requirements of the users.

3.8 Conclusion

This chapter has addressed an anonymous group data transmission problem arising from multi-group RFID systems. The prior works suffer low security due to transmitting group data with plaintext, or low time efficiency due to renewing the keys for encryption. To overcome these drawbacks, we have provided a solution and its enhanced version, namely OBF and OBF+. They use the novel bloom filter where the group data is inserted into the bloom filter in bits thus being overlapped with each other. The OBF shows the feasibility of our idea, and the OBF+ further improves the reliability and the anonymity of the protocol. We also have derived the optimum parameters used in the protocols. We have also implemented the OBF+ in the practical systems and conducted extensive simulations. The results confirm the effectiveness of protocols and the superiority of the OBF+ in terms of reliability and anonymity.

References

1. L. Gao, L. Zhang, M. Ma, Low cost RFID security protocol based on rabin symmetric encryption algorithm. Wireless Personal Commun. **96**(1), 683–696 (2017)
2. H. Yue, C. Zhang, M. Pan, Y. Fang, S. Chen, A time-efficient information collection protocol for large-scale RFID systems, in *2012 Proceedings IEEE INFOCOM* (IEEE, 2012), pp. 2158–2166
3. J. Liu, B. Xiao, S. Chen, F. Zhu, L. Chen, Fast RFID grouping protocols, in *2015 IEEE Conference on Computer Communications (INFOCOM)* (IEEE, 2015), pp. 1948–1956
4. J. Yu, J. Liu, R. Zhang, L. Chen, W. Gong, S. Zhang, Multi-seed group labeling in RFID systems. IEEE Trans. Mobile Comput. **19**(12), 2850–2862 (2019)
5. Y. Qiao, S. Chen, T. Li, S. Chen, Energy-efficient polling protocols in RFID systems, in *ACM MobiHoc* (2011), p. 25
6. J. Yu, J. Liu, R. Zhang et al., Multi-seed group labeling in RFID systems. IEEE TMC **19**(12), 2860–2862 (2019)
7. J. Liu, X. Chen, X. Liu, X. Zhang, X. Wang, L. Chen, On improving write throughput in commodity RFID systems, in *IEEE INFOCOM 2019-IEEE Conference on Computer Communications* (IEEE, 2019), pp. 1522–1530
8. K. Liu, L. Chen, J. Yu, C. Haochen, On batch writing in COTS RFID systems, *IEEE Transactions on Mobile Computing (Early Access)* (2023)
9. J. Yang, B. Liu, H. Yao, Application of chaotic encryption in RFID data transmission security. Int. J. Adv. Network Monitor. Controls **4**(1), 90–96 (2019)
10. J. Yu, W. Gong, J. Liu, L. Chen, K. Wang, On efficient tree-based tag search in large-scale RFID systems. IEEE/ACM ToN **27**(1), 42–55 (2019)
11. J. Yu, L. Chen, R. Zhang, K. Wang, On missing tag detection in multiple-group multiple-region RFID systems. IEEE TMC **16**(5), 1371–1381 (2016)
12. H. Liu, R. Zhang, L. Chen, J. Yu, J. Liu, J. An, On fast and reliable missing event detection protocol for multitagged RFID systems. IEEE IoT J. **7**(10), 10324–10335 (2020)
13. H. Liu, R. Zhang, L. Chen, J. Yu, J. Liu, J. An, Q. Chen, Computation-communication trade-offs for missing multi-tagged item detection in RFID networks, *IEEE Internet of Things Journal* (2021)
14. *EPC radio-frequency identity protocols Class-1 Generation-2 UHF RFID Protocol for communication at 860 MHz - 960 MHz*. EPC, 2.0.1 ed. (2015)

Compact Filter-Based Access Protocol for Multi-Tagged RFID Systems

4

Prior detection protocols are limited to single-tagged RFID systems and would waste considerable time on repeated checks of individual objects in emerging multi-tagged systems, where each object is attached by multiple tags. This inefficiency leaves the challenge of efficient detection in multi-tagged scenarios unaddressed. To bridge the gap, this chapter focuses on detecting missing multi-tagged objects. The key technicality is to build a filter from a subset of tags rather than the entire set, as in previous works, to avoid redundant detection of individual objects and reduce overall detection time. Specifically, we first provide a basic solution based on the Bloom filter which can specify only tags in the chosen subset to participate in final detection. To further improve time efficiency, we propose an advanced protocol that exploits tag ID knowledge and sparsity of slots mapped by the chosen subset to build a more compact compressive filter. Moreover, a composite vector is used to efficiently coordinate tags to report their presence. We further conduct theoretical analysis to determine the optimal protocol parameters and perform extensive simulations to verify the feasibility of the protocols. The results demonstrate that the advanced protocol achieves more than 2x performance gain in terms of time efficiency compared to the Bloom filter-based basic protocol.

Chapter roadmap: The remainder of this chapter is organized as follows. Section 4.1 outlines the motivation for studying the multi-tagged detection RFID systems and summarizes our contributions. Section 4.2 reviews the prior works on missing event detection. The traditional bloom filter used in both marking and detection is described in Sect. 4.4. In Sect. 4.5, we introduce our marking method with a compressed filter and the detection method based on multiple seed detecting. Section 4.6 discusses the simulation results of the proposed protocols in different scenes. Finally, we conclude this chapter in Sect. 4.7.

© The Author(s), under exclusive license to Springer Nature Switzerland AG 2026
R. Zhang and H. Liu, *RFID Applications*, Synthesis Lectures on Communications,
https://doi.org/10.1007/978-3-031-93034-8_4

4.1 Introduction

This chapter focuses on a variation on missing event detection problems different from prior works, motivated by the emerging deployment of multi-tagged RFID systems where each object in the coverage is attached with multiple tags. In this chapter, we use filters to mark a subset of the entire tags and conduct the marked tags to access the reader. The main contributions of this chapter are articulated as follows.

- First, we provide an efficient solution to the missing event detection problem in multi-tagged RFID systems, named basic protocol. In the first phase, we leverage the Bloom filter to represent the chosen tags so that they can pass the membership test while the others are sifted out. A virtual Bloom filter is constructed from responses of the tags in the second phase, enabling missing tag detection.
- Second, we design an advanced protocol that is more time efficient. Exploiting properties of full knowledge of tags' IDs and sparsity of slots mapped by the chosen tags compared with the others, we propose a compressive filter that only needs one hashing operation for a tag but can achieve better marking efficiency than the Bloom filter. A composite vector built from multiple mappings of the marked tags is then used for the detection.
- Third, we investigate the performance of the proposed protocols both theoretically and experimentally. We derive optimum parameters used in the protocols that minimize communication overhead under the constraint of required detection reliability. On the other hand, extensive simulation results verify the effectiveness of both protocols on missing event detection and show that the advanced protocol achieves a time efficiency gain of at least 2x over the Bloom filter-based basic one.

4.2 Related Work

Missing tag detection plays a crucial role in RFID-enabled applications since it can monitor the state (normal or broken) of tags and quickly detect illegal movement of objects in work regions such as misplacement and theft. The works on missing tag detection could be separated into two categories: probabilistic [1, 2, 2–7] or deterministic protocols [8–10].

Probabilistic protocols detect a missing tag event with a predefined probability. Tan et al. initiate the study of probabilistic detection and propose a solution called Trusted Reader Protocol (TRP) in [1]. TRP detects a missing tag event by comparing the pre-computed slots with those picked by the tags in the population. If an expected singleton slot turns out to be an empty slot, then the missing event is detected. Follow-up works [2, 3] employ multiple seeds to increase the probability of the singleton slot, which reduces the useless empty and collision slots and thus achieves better performance. RUN [4] and BMTD [5] are proposed to address the influence of unknown tags. Yu et al. [6] design a suit of detection protocols for multi-categories and multi-region RFID systems and study how to detect missing tags by using COTS RFID devices [7].

Deterministic protocols, on the other hand, are able to exactly identify which tags are absent. Li et al. develop a series of deterministic protocols in [8] to reduce the radio collision by reconciling collision slots and finally iron out a bit-level tag identification method by iteratively deactivating the tags of which the presence has been verified. Subsequently, Zhang et al. propose identification protocols in [9] which store and compare the bitmap of tag responses in all rounds and observe the change among the corresponding bits among all bitmaps to determine the present and absent tags. However, how to configure the protocol parameters is not theoretically analyzed. More recently, Liu et al. [10] enhanced the work by reconciling both 2-collision and 3-collision slots and filtering the empty and unreconcilable collision slots to improve time efficiency.

We would like to emphasize that none of the prior works is designed to detect missing events in a multi-tagged RFID system. In this scenario, all existing missing tag detection protocols cannot work effectively, because they have to detect all tags whose IDs are recorded in the reader, wasting too much time. In contrast, our work chooses a subset of these tags for detection, avoiding repeated checks of one object and its interferences with the other tags. Moreover, this chapter exploits tag knowability and slot sparsity jointly to improve time efficiency, which completely differs our work from the existing ones.

4.3 System Model and Problem Formulation

4.3.1 System Model

We consider an RFID system of one reader[1] and a large number of tags where each physical object is attached by multiple tags [12, 13]. The reader is connected via high-speed channels with a back-end server of powerful computing capability. We regard the server and the reader as a single entity called *the reader* for simplicity [14, 15]. Generally, each tag has a unique ID and user-defined memory to achieve storage of the user-defined data while capable of performing certain computations like hashing functions. Moreover, we assume that the reader has the IDs of all tags in the system.

The downlink (i.e., reader-to-tags) and uplink (i.e., tags-to-reader) communications experience different slot duration: (1) 96-bit downlink slot duration from the reader to tags; (2) 1-bit response slot from tags to the reader. We denote T_{id} and T_{short} as the length of a downlink slot and response slot, respectively. For an arbitrate response slot, there are three types of slot states. If no tag relies on this slot, it is called an empty slot; if a single tag replies,

---

[1] For multiple readers, we can treat them as a single virtual reader as in [6, 11]. Specifically, the back-end server calculates all the parameters constructs the filter vectors, and sends them to all readers such that the readers broadcast the same parameters and filters to the tags. Consequently, the back-end server can synchronize the readers and we can logically consider them as a whole.

it is called a singleton slot; if multiple tags respond simultaneously, it is called a collision slot. The latter two states are referred to as non-empty slots.

4.3.2 Problem Formulation

In this chapter, we are interested in detecting missing object events in a multi-tagged RFID system where n tags monitor g objects, and each object is tagged by multiple tags, i.e., $g < n$. Let m_a denote the number of missing objects, a missing event denotes the event that m_a exceeds a threshold M_a. Let P_d define the probability that the reader can find a missing event, we formulate the optimum missing event detection problem as follows: *The missing multi-tagged object detection problem is to devise an algorithm of minimum execution time to find a missing event with probability $P_d \geq \alpha$ when $m_a \geq M_a$, where α is the required detection reliability.* Given the required probability, the key performance metric is communication overhead between the reader and tags spent in completing the detection task. In this chapter, the communication overhead means the execution time.

We would like to emphasize the main difference between the problem in this chapter and those in the prior works: In our problem, one missing object leads to multiple tags absent from the interrogation of the reader. Instead, an object and its attached tag are injective in the prior work. This difference makes the algorithm design in this chapter completely different, which can be interpreted as follows: If a tag is absent from the interrogation of the reader, the corresponding attached object can be regarded as missing in the prior work. This, however, does not hold for the multi-tagged system here. In the new scenario, the reader learns a missing object only when all its attached tags are absent. If we still use the prior algorithms to deal with the new problem, all tags on an object would respond to the interrogation, leading to severe interference and thus considerably degrading time efficiency.

Take an example to explain the difference. Consider 10, 000 objects, there will be then 10, 000 tags detected by the reader in an injective RFID system. Yet, the number will soar to 30, 000 in a multi-tagged system where each object is attached by 3 tags if the existing algorithms are used, sharply increasing communication overhead. This urges us to investigate the following problem: can we design detection algorithms that can achieve the required detection reliability by interrogating only part of the tags in the system? We shall answer this question later in this chapter with a comprehensive investigation. Table 4.1 summarizes main notations used in the chapter.

4.3.3 Design Rational

Recall the missing multi-tagged object detection problem, an object is missing if all of its attached tags are absent, but the absence of one tag indicates the potential missing object. Consequently, it is adequate to first probe one of the tags on an object instead of all for

4.3 System Model and Problem Formulation

Table 4.1 Main parameter notation

Symbols	Description
$\mathcal{G}_A$	Set of representative tags
$\mathcal{G}_B$	Set of pending tags
n	Number of tags in our system
g	Number of objects in our work region
P_d	Achieved detection probability
α	Requirement of detection probability
m_a	Number of actual missing representative tags
M_a	Least number of missing representative tags to satisfy detection requirement
f_1	Length of filter in marking via bloom filter
k_1	Number of mapping in marking via bloom filter
P_{fp_1}	Probability of false positives in marking
f_2	Length of filter in detection via bloom filter
k_2	Number of mapping in detection via bloom filter
P_{fp_2}	Probability of false positives in detection
λ	Marking Efficiency in the advanced protocol
f_d	The response frame length in the second phase of the advanced protocol

missing object event detection. If the probed tag is present, the tagged object must still be located in the coverage of the RFID system and we do not need to interrogate the other tags on this object, which reduces communication overhead. Otherwise, we would further poll the Big tags on the object, and a missing object can be found if all of them are absent. Since the percentage of missing objects is usually small, the idea above can improve time efficiency.

Following the guideline, we randomly choose a tag from each object, which is referred to as **representative tag**. These g tags constitute the representative tag set defined as $\mathcal{G}_A = \{tag_1, tag_2, ..., tag_g\}$ where tag_i is a tag on the object i for $1 \leq i \leq g$. The set of the remaining tags named **pending tags** is denoted by $\mathcal{G}_B$. We then are interested in interrogating the representative tags to detect potential missing object events. Yet the pending tags in $\mathcal{G}_B$ would cause severe interference to representative tag detection. Therefore, an efficient scheme should be able to eliminate this negative impact.

In this chapter, we design two-phase protocols to address the problem: (1) Marking phase: The task of Phase 1 is to mark the representative tags for further detection while depressing the pending tags to abate their interference. The key to answering this question lies in designing a filter that is able to filter out the pending tags while ensuring all representative tags pass the test; (2) Detecting phase: The reader then conducts missing object event detection in Phase 2 by interrogating the remaining tags after the execution of Phase 1. Therefore, we should ensure the efficiency of the two phases so that the overall time cost can be minimized. To this end, we propose two approaches. Note that a filter is an indicator vector with a certain number of elements each being either '0' or '1', and a position in an offline built filter corresponds to the slot in the same sequence of a frame during the online execution.

Basic approach: Bloom filter-based algorithm. Bloom filter is a space-efficient probabilistic data structure for representing a set and supporting set membership queries. Its property can meet the design requirements analyzed above. Specifically, the reader first constructs a bloom filter with the optimum parameters by encoding each tag in $\mathcal{G}_A$ and transmits parameters and the filter to all tags. On the tag side, each tag uses the hash functions and the received parameters to map itself to several positions in the received filter. If all the value of these positions is '1', the tag knows it is a representative tag and will participate in the detection in Phase 2. Otherwise, the tag is a pending tag and should turn to sleep and wait for the next activation command. This method is a direct application of a bloom filter to achieve the marking task. After the marking phase, the reader detects missing tags by constructing a virtual bloom filter from the responses of the active tags. Since the reader can predict slot states, it can find a tag missing if there exists at least one of its mapped slots which is supposed to be busy but turns out empty.

Advanced approach: Compressive filter-based algorithm. Bloom filter can effectively complete the marking task, yet its performance is hindered by the tradeoff between filter length (i.e., frame size) and false positive ratio that tags in $\mathcal{G}_B$ are mistakenly marked with a certain probability. Specifically, reducing the false positive ratio is at the price of a longer filter. Especially, when $|\mathcal{G}_B|$ is considerably larger than $|\mathcal{G}_A|$, we should accordingly increase filter length to reduce the false positive ratio, and a higher false positive ratio leads to severe interference to the representative tags, otherwise.

To tackle the drawback of the basic approach, we develop a new filter that only needs one hash function rather than multiple ones in the Bloom filter but can achieve better performance. First, the reader employs one hash function to construct a filter where all positions are initialized to '0' and only those mapped by tag(s) from $\mathcal{G}_A$ are set to '1'. Such a filter can mark tags in $\mathcal{G}_A$ and ask them to participate the second phase. Second, to reduce the time cost spent on the filter transmission, we explore the sparsity of '1' in the filter to compress its size.

Specifically, the elements '0' in the filter are in the majority, and its proportion increases with the filter size and the difference of $\mathcal{G}_B$ and $\mathcal{G}_A$. Moreover, the filter performs as a binary test, it is thus adequate to inform the tags of the positions of '1' in the filter. Motivated by these observations, we design such a compressive algorithm that consecutive zeros between any two '1' in the filter are replaced by a binary bit sequence of fixed size. It is required that the denary value of the bit sequence is equal to the number of consecutive zeros, which can be used to indicate the positions of '1' in the original filter. Through the optimum parameter configuration, the compressive filter can be significantly more complex than the original one.

In the second phase, since a missing tag will be found when it is mapped to a singleton slot, we aim to improve communication efficiency by changing non-singleton slots into singleton slots. On the reader side, it first offline maps each representative tag independently via different seeds and builds a composite vector by picking all singleton slots from the multiple mapping. It then broadcasts parameters including the vector, its size, and the seeds. At the tag side, each tag maps to one position of the vector using one seed and should respond if finding the mapping slot is a singleton. From the responses of tags, the reader can check whether a representative tag is missing and decide whether to poll the remaining tags in the corresponding object to verify its existence.

In what follows, we elaborate on the basic approach and the advanced one in subsequence.

4.4 Basic Approach: Bloom Filter-Based Protocol

In the basic approach, downlink and uplink bloom filters are built in the two phases for missing event detection, respectively. In Phase 1, the reader first constructs a bloom filter to mark representative tags by encoding each tag in $\mathcal{G}_A$ according to the derived parameters and transmits the parameters and the constructed bloom filter to tags. Tags conduct a membership test by checking the value of its mapping positions in the received filter. The details of the method will be described as follows. In Phase 2, the reader interrogates the remaining active tags with another suit of the derived parameters. Each tag should reply in its mapping slots, and a virtual bloom filter can be constructed from the responses of all tags at the reader side for missing tag detection.

4.4.1 Protocol Description

The basic protocol consists of two phases: The marking phase and the detection phase, as described below.

(1) Marking phase: In the beginning, the reader samples tags to participate in this process. To achieve sampling probability of p_1, the reader broadcasts parameters of length f_{sample}, seed s_{sample} and threshold $Th_1 = \lceil p_1 f_{sample} \rceil$. Each tag hashes to $[0, f_{sample})$ with s_{sample}. If the result is smaller than Th_1, it will take part in this process, and keep sleep, otherwise.

The rest of the first phase can be executed in multiple rounds, which is decided by the parameter configuration to be discussed in Sect. 4.4.2. Recall that the objective of this phase is to filter out the pending tags in $\mathcal{G}_B$. We consider the ith round mark of $\mathcal{G}_A$, $1 \leq i \leq R_1$, where R_1 is the number of executed rounds. Let B_i be the number of the still active pending tags at the beginning of this round.

The Reader offline constructs a f_1-bit bloom filter BV_i by mapping each tag ID in $\mathcal{G}_A$ to k_1 positions under seed s_i and set their value to '1'. Then, the reader broadcasts the parameters and BV_i. Each unmarked tag uses the same parameters to map itself to k_1 positions as the reader does. If the tag finds all the mapped k_1 bits in BV_i are ones, it passes the filter and waits for the detection in the second phase. Otherwise, it will keep asleep and cannot take part in the rest of the protocol. The Bloom filter has no false negative, i.e., tags in $\mathcal{G}_A$ must pass the test, but it suffers from false positive: A tag in $\mathcal{G}_B$ may also pass the check. We denote by g_i the number of the tags filtered out in this round. After all R_1 rounds, there will be $B_{R_1} - g_{R_1}$ active tags in $\mathcal{G}_B$ which will access to the second phase.

(2) Detection phase: This phase aims to detect potential missing representative tags in $\mathcal{G}_A$ with the presence of $B_{R_1} - g_{R_1}$ active tags of $\mathcal{G}_B$. Similar to the first phase, the reader also first samples the remaining tags with a sampling probability of p_2 and threshold $Th_2 = \lceil p_2 f_{sample} \rceil$. The rest of the second phase is executed in multiple rounds, which is decided by the parameter configuration to be discussed in Sect. 4.4.2.

Denote by R_2 the number of rounds in this phase. Consider an arbitrate round i, different from the first phase, a Bloom filter will be built from the responses of the tags, which is used by the reader to check the existence of each tag. To this end, the reader broadcasts the parameters including filter size f_2, the number of hush functions k_2, and seed s_2^*. Each tag then maps itself to k_2 slots and will reply in these slots. On the reader side, it can build a bloom filter by setting positions corresponding to busy slots to '1'. As the reader knows all IDs, it can predict every slot state and can thus detect a missing tag if there exists at least one '0' at its mapped k_2 positions.

Although there are false positives and the interference of some pending tags, we could configure parameters used in the protocol so that the required detection reliability can be met within the minimum communication overhead. The analysis will be introduced in Sect. 4.4.2.

4.4.2 Parameter Optimization

The execution time of the basic protocol mainly consists of two parts: the communication cost in the marking phase and that spent on the detection.

(1) We start with the analysis of the first part. The execution time of the marking phase could be expressed as

$$T_m = T_{m\_ini} + f_1 R_1 \frac{T_{id}}{96}, \tag{4.1}$$

where T_{m_ini} is the constant time cost of the parameter transmission. The goal is thus to minimize $f_1 R_1 \frac{T_{id}}{96}$.

It is known that the false positives of the bloom filter are

$$P_{fp_1} = \left[1 - \left(1 - \frac{1}{f_1}\right)^{k_1 A}\right]^{k_1} \approx \left(1 - e^{-\frac{k_1 A}{f_1}}\right)^{k_1}, \tag{4.2}$$

where $A = A_{orig} p_1$ is the number of tags passing the sampling in $\mathcal{G}_A$, k_1 is the number of hash functions and f_1 is the length of the Bloom filter (i.e., frame size). Consider an arbitrary round, if the k_1 slots mapped by a tag in $\mathcal{G}_B$ are the same as those in $\mathcal{G}_A$, then it cannot be

4.4 Basic Approach: Bloom Filter-Based Protocol

filtered out in this round. The probability of this event is (4.2). Therefore, the probability that a tag in $\mathcal{G}_B$ remains active after the marking phase can be written as

$$P_{fp_1}^{R_1} = \left(1 - e^{-\frac{k_1 A}{f_1}}\right)^{k_1 R_1}, \tag{4.3}$$

where R_1 is the number of executing rounds.

We calculate the first order of differential function and obtain the minimum value of $P_{fp_1}^{R_1}$ is $\left(\frac{1}{2}\right)^{\frac{f_1 R_1}{A}} \ln 2 \approx 0.6185^{\frac{f_1 R_1}{A}}$ when $k_1 = \frac{f_1}{A} \ln 2$. Therefore, the key is to minimizing $f_1 R_1$. Due to the fact that a smaller $f_1 R_1$ results in more active pending tags and more interferences to the detection phase, we thus jointly minimize the cost with the second phase.

(2) We define the cost of execution time in the detection phase as T_d

$$T_d = T_{d\_ini} + f_2 R_2 T_{short}. \tag{4.4}$$

Similarly, we should minimize $f_2 R_2$ for time cost optimization with the constraint of the detection reliability. To this end, we first calculate the probability of false positives in the detection phase, as expressed in the below:

$$P_{fp_2}^{R_2} = \left(1 - e^{-\frac{k_2 A'}{f_2}}\right)^{k_2 R_2}, \tag{4.5}$$

where f_2 is the frame length, k_2 is the number of mappings (i.e., the number of hash functions) in a frame, and A' is the number of tags responding to the interrogation. Denote by A_r the number of the remaining tags after the first phase and m is the number of missing tags, then $A' = (A_r - m)p_2$. Similar with $P_{fp_1}^{R_1}$, we have the minimum $P_{fp_2}^{R_2}$:

$$P_{fp_2}^{R_2} = 0.6185^{\frac{f_2 R_2}{A'}}. \tag{4.6}$$

We denote by P_d the probability that a missing event could be detected in $\mathcal{G}_A$. As we should detect the missing event when $m_a \geq M_a$, P_d could be derived as

$$P_d = 1 - [1 - p_1 + p_1(1 - p_2 + p_2 P_{fp_2}^{R_2})]^{M_a}, \tag{4.7}$$

where p_1, p_2 are sampling ratios in the two phases, respectively. In order to meet system requirement in detection, P_d should be greater than α, then we have

$$P_{fp_2}^{R_2} \leq \frac{\frac{(1-\alpha)^{\frac{1}{M_a}} + p_1 - 1}{p_1} + p_2 - 1}{p_2}. \tag{4.8}$$

It is required that

$$p_1 p_2 > 1 - (1 - \alpha)^{\frac{1}{M_a}}. \tag{4.9}$$

As it is adequate to set $P_d = \alpha$, we have

$$f_2 R_2 = \frac{A'}{-(\ln(2))^2} \cdot \left[\ln\left(\frac{(1-\alpha)^{\frac{1}{M_a}} + p_1 - 1}{p_1} + p_2 - 1\right) - \ln(p_2) \right]. \quad (4.10)$$

Recall that A' is the number of tags responding to the interrogation including partial tags of $\mathcal{G}_A$ and a few of $\mathcal{G}_B$, as it is enough to find missing tags when $m_a \geq M_a$, we can rewrite A' for the parameter settings as

$$A' = (A + BP_{fp_1}^{R_1} - M_a)p_2, \quad (4.11)$$

where $B = B_1 = B_{orig} p_1$. Substituting (4.6) and (4.11) into (4.10), we have

$$f_2 R_2 = \frac{\ln\left(\frac{(1-\alpha)^{\frac{1}{M_a}} + p_1 - 1}{p_1} + p_2 - 1\right) - \ln(p_2)}{-(\ln(2))^2} \cdot \left(A + B 0.6185^{\frac{f_1 R_1}{A}} - M_a\right) p_2. \quad (4.12)$$

(3) From (4.12), we can observe that T_d increases with the decrease of $f_1 R_1$ that is determined by the first phase. Define the overall time cost of the basic protocol as T_{whole}, we have

$$T_{whole} = T_{m\_ini} + T_{d\_ini} + f_1 R_1 \frac{T_{id}}{96} + f_2 R_2 T_{short}. \quad (4.13)$$

Since T_{g_ini} and T_{d_ini} are constants and too small compared with the other parts. Hence, we ignore them in the subsequent optimization. The overall cost is simplified as

$$\hat{T} = f_1 R_1 \frac{T_{id}}{96} + f_2 R_2 T_{short}$$

$$= \frac{\ln\left(\frac{(1-\alpha)^{\frac{1}{M_a}} + p_1 - 1}{p_1} + p_2 - 1\right) - \ln(p_2)}{-(\ln(2))^2} T_{short} \cdot \left(A + B 0.6185^{\frac{f_1 R_1}{A}} - M_a\right) p_2 + \frac{T_{id}}{96} f_1 R_1. \quad (4.14)$$

Denote $u = f_1 R_1$, we derive the differential of $\hat{T}$ with u:

$$\frac{\partial \hat{T}}{\partial u} = \left[\ln\left(\frac{(1-\alpha)^{\frac{1}{M_a}} + p_1 - 1}{p_1} + p_2 - 1\right) - \ln(p_2)\right] T_{short} \cdot \frac{B p_2 0.6185^{\frac{u}{A}}}{A} + \frac{T_{id}}{96}. \quad (4.15)$$

Let $\frac{\partial \hat{T}}{\partial u} = 0$, we could get the minimum overall when

$$u = -\frac{A}{\left(\ln 2\right)^2} \times \ln\left(\frac{-\frac{T_{id}}{96} A}{T_{short} B p_2 \left(\ln\left(\frac{(1-\alpha)^{\frac{1}{M_a}} + p_1 - 1}{p_1} + p_2 - 1\right) - \ln p_2\right)}\right). \quad (4.16)$$

4.5 Advanced Approach: Compressive Filter-Based Protocol

Parameter configuration: Given the sampling ratios p_1 and p_2 meeting (4.9), the value of f_1 and R_1 can be chosen so that (4.16) holds. Once they are fixed, we can get f_2 and R_2 following (4.12). Finally, the optimal parameters can be configured for the basic protocol.

4.5 Advanced Approach: Compressive Filter-Based Protocol

Bloom filter can effectively complete the marking task, yet its performance is hindered by the tradeoff between filter length (i.e., frame size) and false positive ratio that tags in $\mathcal{G}_B$ are mistakenly marked with a certain probability. Specifically, reducing the false positive ratio is at the price of a longer filter. Especially, when $|\mathcal{G}_B|$ is considerably larger than $|\mathcal{G}_A|$, we should accordingly increase filter length to reduce the false positive ratio, and a higher false positive ratio leads to severe interference to the representative tags, otherwise.

To tackle the drawback of the basic protocol, we develop an advanced protocol containing a new filter for the marking phase that only needs one hash function rather than multiple in Bloom filter but can achieve better marking performance, and a composite filter picking all singleton slots from multiple mappings. *The improvement in the first phase results from two aspects: The knowledge of IDs of all tags, and the sparsity of the original vector.* The first one enables the reader to encode the mappings of both representative and pending tags instead of only the former in the basic protocol, making the filter more informative. The second one makes compression of the filter possible reducing communication costs.

4.5.1 Protocol Description

The advanced protocol also consists of two phases: The marking phase and the detection phase. In the first phase, we use one hash function to encode mappings of all tags and exploit the sparsity of '1' to build a compressive filter to mark representative tags. In the second phase, we pick singleton slots from multiple random mappings of a tag to build a composite filter informing a remaining active tag after the first phase of its response slot and conduct the detection. Note that the position of a filter and a slot of a frame is injective.

(1) Marking phase: In the marking phase, the reader first samples the tags with ratio of p_1. Then, the marking phase works in multiple rounds. Consider an arbitrary round i, the reader offline employs one hash function to encode all unmarked tags to an f_i-bit vector where all positions are initialized to '0'. Since the reader knows IDs of all tags, it can predict A-homogeneous positions that are mapped only by tag(s) of $\mathcal{G}_A$, B-homogeneous positions that are mapped only by tag(s) of $\mathcal{G}_B$, heterogeneous positions that are mapped by tags of $\mathcal{G}_A$ and $\mathcal{G}_B$, and empty positions. Consequently, the reader only sets the A-homogeneous positions of the vector to '1' instead of both homogeneous and heterogeneous positions in

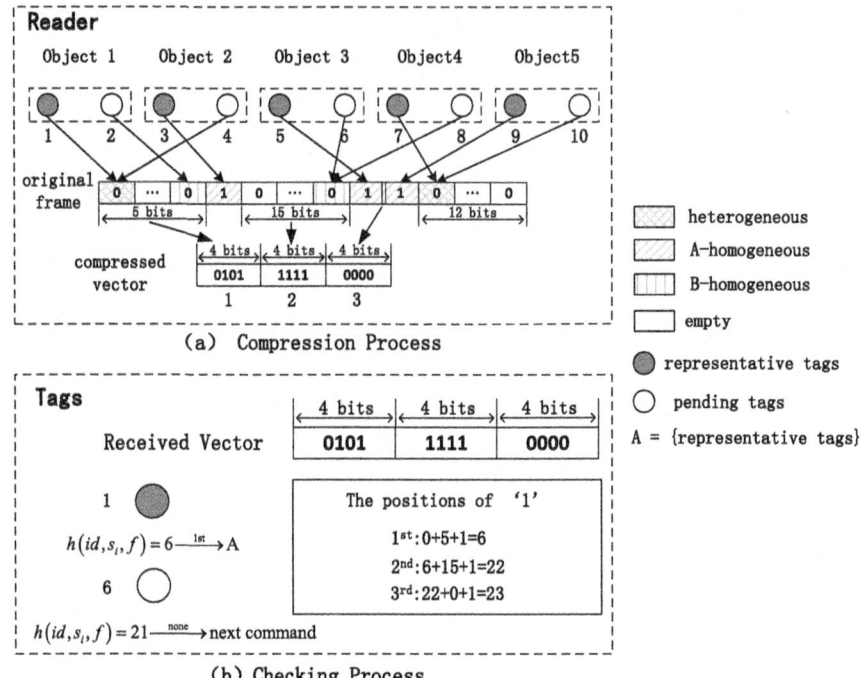

Fig. 4.1 An illustration of a compressive filter and the checking process at tag side

the basic protocol. Note that a position in an offline built vector corresponds to the slot in the same sequence of a frame during the online execution.

Let's take Fig. 4.1a as a toy example. The 1st position is heterogeneous because it is mapped by tag 1 in set $\mathcal{G}_A$ and tag 4 in set $\mathcal{G}_B$. The 21st position is also set to 0 since it is a B-homogeneous position mapped by tag 6 and tag 8 of set $\mathcal{G}_B$. On the contrary, the 6th and 22nd positions are A-homogeneous since they are mapped by tags in group $\mathcal{G}_A$. Following the rule, we can build the original vector as '0000_0100_0000_0000_0000_0110_0000_0000_000'.

After the original vector is built, we start to compress it, which is motivated by the sparsity of '1' as shown in Fig. 4.1a. We exploit the distance between two '1' to indicate the positions of '1' in the vector. Because the distance is usually short, the vector length can be significantly reduced. Specifically, the reader replaces each segment of consecutive zeros between '1' by the number of consecutive zeros in this segment. To this end, the reader first finds the longest segment of consecutive zeros in the original vector and records the length of zeros as L_i^{max}. Second, each segment of consecutive zeros is converted to a binary sequence of $l_i = \lceil \log_2 \left(L_i^{max} + 1 \right) \rceil$ bits whose decimal value is equal to the number of consecutive zeros, and the compressive filter is finally constructed. If the compressive filter is longer than 96 bits, the reader can divide it into parts and transmit each part in T_{id}.

4.5 Advanced Approach: Compressive Filter-Based Protocol

For instance in Fig. 4.1a, the longest segment of 15 zeros is converted to the number 15, which is compressed from 15 bits to 4 bits, and the other segments are also represented as 4-bit sequences. Consequently, the reader can get a 12-bit compressive filter compressed from the 35-bit original vector. The compression ratio is $12/35 \approx 0.34$.

The reader then broadcasts parameters including original vector size f_i, the segment size l_i, and seed s_i. We will analyze how to set the parameters in Sect. 4.5.2. The reader also sends the compressive filter to tags. At the tag side, after receiving the filter, it calculates the decimal value of each l_i-bit segment starting from the head of the filter and outputs the same number of consecutive zeros. Repeat this for all segments, a tag can learn all positions of value '1' among $[1, f_j]$. It then can directly check from the compressed filter whether it is a representative tag. Specifically, the tag hashes itself to a slot among $[1, f_j]$. It then subtracts the sum shown in Fig. 4.1b from its hash value until the result is non-positive. It can be marked as a representative tag if the result is zero. Otherwise, it waits for the following marking round. Note that it means two consecutive '1' that the decimal value of a compressed segment is 0. Moreover, the length of the reconstructed vector may be smaller than f_i because the consecutive zeros at the end of the original vector are omitted to save time cost. The tag just needs to fill with several zeros at the end of the reconstructed vector to reach f_i. After multi-round execution, all sampled representative tags can be marked and access the detection phase, while the others keep silent.

Let's take Fig. 4.1b as an example to illustrate the decompression process at the tag side. From the received compressive filter, tag 3 can learn that there are 5 zeros until the first '1', matching with its mapping, it can thus be marked. In contrast, tag 6 mapped to the 21st slot finds the value at the 21st position of the original vector is '0', which can be inferred from 15 zeros between the 1st and 2nd '1'. It thus knows that it should keep silent in the rest of the protocol.

(2) Detection phase: In this phase, the reader first samples the tags marked in the first phase with the ratio of p_2. The reader then constructs a composite vector from multiple mappings of the sampled tags. Define the composite vector length as f_d and seed sequence $\{s_1, s_2, ..., s_l\}$. We will analyze how to set the parameters in Sect. 4.5.2. The reader maps a tag to $H(id, s_j, f_d)$th position of the jth vector in the jth mapping where $1 \leq j \leq l$. After l mappings of all tags, the reader can obtain l vectors and use them to composite a vector storing indexes of seeds that contribute to singleton slots. Specifically, the f_d-bit composite vector is initialized to null. For each of its positions i, the reader picks a seed that makes one of the ith positions in the obtained l vectors singleton, for example, s_j, and sets the ith position in the composite vector to j. Repeating these operations for all f_d positions, the reader can obtain the expected composite vector.

After the offline construction of the composite vector, the reader broadcasts the vector length f_d, seed sequence $\{s_1, s_2, ..., s_l\}$ and the composite vector. And the reader sends another interrogation command to ask the qualified tag to respond, subsequently. At the tag side, for each slot, each tag uses a seed to map itself to a position of the vector and checks whether the sequence of the position in the vector is equal to the slot in the frame and whether the seed index in this position of the vector is equal to the seed used in this mapping. If both of them hold, the tag will respond in this slot. Otherwise, it uses another seed and repeats the above operations. On the reader side, the reader can compare the observed slot states with the predicted ones. It can detect a missing tag if a predicted singleton slot turns out to be empty.

4.5.2 Parameter Setting

We here introduce how to set parameters so that the detection reliability can be met and the communication cost can be minimized. To make the analysis feasible, we separately analyze the communication cost of the two phases.

(1) Optimum parameters for the marking phase: In an arbitrary round i of this phase, the objective is to maximize the marking efficiency λ_i: The ratio of the number ϕ_i of sampled representative tags in $\mathcal{G}_A$ marked in this round to the execution time t_i of this round. It implies that more tags can be marked in a unit of time when λ increases. Let f_i^c define the compressive filter length in this round, we have

$$\lambda_i = \frac{\phi_i}{t_i} = \frac{\phi_i}{\frac{f_i^c}{96} T_{id}}. \tag{4.17}$$

As ϕ_i and f_i^c depend on f_i, the key is to find the optimum f_i.

Let n_i be the number of sampled representative tags unmarked at the beginning of the round, and when all sampled representative tags are marked after I rounds, n_I equals to the number of sampled pending tags in $\mathcal{G}_B$ in this phase. Denote by ϕ_i' the number of sampled representative tags unmarked at the beginning of the round, we have

$$n_{i+1} = n_i - \phi_i,$$
$$\phi_{i+1}' = \phi_i' - \phi_i. \tag{4.18}$$

Since the protocol is probabilistic, we derive the expected value of ϕ_i, and the result is stated in the following lemma.

Lemma 4 *Given the original vector size f_i at the ith round, the expected number of sampled representative tags marked in this round should be*

$$\phi_i = \phi_i'\left(1 - \frac{1}{f_i}\right)^{n_i - \phi_i'}. \tag{4.19}$$

Proof We first study the event that j sampled representative tags are mapped to the same slot. Its probability, defined as P_A^i, consists of there parts: The probability of an arbitrary

4.5 Advanced Approach: Compressive Filter-Based Protocol

slot mapped by j tags which is $(\frac{1}{f_i})^j (1 - \frac{1}{f_i})^{n_i-j}$, and $\binom{n_i}{j}$ kinds of combination of j tags, and the probability of j tags being representative tags which is equal to $\mathcal{G}_A$ is $\binom{\phi'_i}{j}/\binom{n_i}{j}$. It thus holds that

$$P_A^i = \binom{\phi'_i}{j}\left(\frac{1}{f_i}\right)^j\left(1 - \frac{1}{f_i}\right)^{n_i-j}. \tag{4.20}$$

Hence, the expected number of sampled tags in group $\mathcal{G}_A$ mapped to a slot is $\sum_{j=0}^{\phi'_i} j \binom{\phi'_i}{j}$ $\left(\frac{1}{f_i}\right)^j \left(1 - \frac{1}{f_i}\right)^{n_i-j}$, and the number of the sampled tags in group $\mathcal{G}_A$ marked by the compressive vector could be written as

$$\phi_i = f_i \sum_{j=0}^{\phi'_i} j \binom{\phi'_i}{j}\left(\frac{1}{f_i}\right)^j\left(1 - \frac{1}{f_i}\right)^{n_i-j}.$$

After algebraic operations, the lemma can be proven.

From the construction of the compressive filter, we can find the following relation between f_i^c and f_i

$$f_i^c = f_i(1 - \frac{1}{f_i})^{n_i - \phi'_i}(1 - (1 - \frac{1}{f_i})^{\phi'_i}) \times \log_2\left(\frac{1}{(1 - \frac{1}{f_i})^{n_i - \phi'_i}(1 - (1 - \frac{1}{f_i})^{\phi'_i})} + 1\right), \tag{4.21}$$

where the multiplicators at the two sides of the multiplication sign are the expected number of A-homogeneous positions and the average length of consecutive zeros in the original vector, i.e., l_i, respectively. The relation among n_i, ϕ_i and ϕ'_i also satisfies (4.18). Substituting (4.21) into (4.17), we can approximately have

$$\lambda_i = \frac{96}{T_{id}} \frac{\phi'_i}{f_i\left(1 - \left(1 - \frac{1}{f_i}\right)^{\phi'_i}\right)} \times \frac{1}{\log_2\left(\frac{1}{(1-\frac{1}{f_i})^{n_i-\phi'_i}\left(1-(1-\frac{1}{f_i})^{\phi'_i}\right)} + 1\right)}. \tag{4.22}$$

To accelerate the mark phase, we should select an optimum f_i that maximizes the marking efficiency λ_i. To this end, we conduct theoretical analysis and provide a upper bound for the optimum f_i, which is stated in the following theorem.

Theorem 6 *Given n_i and ϕ'_i that are known at the beginning of round i, the optimum f_i falls in $[1, \frac{n_i^2}{n_i - 0.5\phi'_i}]$.*

Proof As it is unfeasible to directly derive optimum f_i from (4.22), we derive an upper bound of f_i and prove that λ_i is a decreasing function with respect to f_i when f_i exceeds this upper bound. As a result, the optimum f_i maximizing λ_i can be found between 1 and this upper bound.

Let $b = 1 - (1 - \frac{1}{f_i})^{\phi'_i}$. We can write

$$\frac{1}{\lambda_i} = \frac{T_{id}}{96\phi'_i} f_i b \log\left(1 + \frac{1}{(1-b)^{\frac{n_i}{\phi'_i}-1} b}\right).$$

We can check that $\frac{1}{(1-b)^{\frac{n_i}{\phi'_i}-1} b}$ is decreasing in b for $0 \le b \le \frac{\phi'_i}{n_i}$. Hence $\log\left(1 + \frac{1}{(1-b)^{\frac{n_i}{\phi'_i}-1} b}\right)$ is decreasing in b. Note that it easy to check that b also decreases with f_i, $\log\left(1 + \frac{1}{(1-b)^{\frac{n_i}{\phi'_i}-1} b}\right)$ is thus increasing in f_i. On the other hand, regard $y = f_i b$ as a function of f_i, we can derive that

$$y' = 1 - \left(\frac{f-1}{f}\right)^{\phi'_i} \left(1 + \frac{\phi'_i}{f_i - 1}\right) > 0. \tag{4.23}$$

Therefore, $\frac{1}{\lambda_i}$ is increasing in f_i when $0 \le b \le \frac{\phi'_i}{n_i}$. To establish the inequalities, f_i should satisfy that

$$f_i \ge \frac{1}{1 - (1 - \frac{\phi'_i}{n_i})^{\frac{1}{\phi'_i}}}. \tag{4.24}$$

By applying Taylor series $1 - zx < (1-x)^z < 1 - zx + 0.5zx^2$, we have

$$\frac{n_i - 0.5\phi'_i}{n_i^2} < 1 - \left(1 - \frac{\phi'_i}{n_i}\right)^{\frac{1}{\phi'_i}} < \frac{1}{n_i}.$$

Hence it is adequate to guarantee that $\frac{1}{\lambda_i}$ is increasing in f_i for $f_i \ge \frac{n_i^2}{n_i - 0.5\phi'_i}$. Consequently, λ_i is decreasing when $f_i \ge \frac{n_i^2}{n_i - 0.5\phi'_i}$. It thus suffices to search f_i to find its optimum value until λ_i starts to decrease. The theorem follows from here.

4.5 Advanced Approach: Compressive Filter-Based Protocol

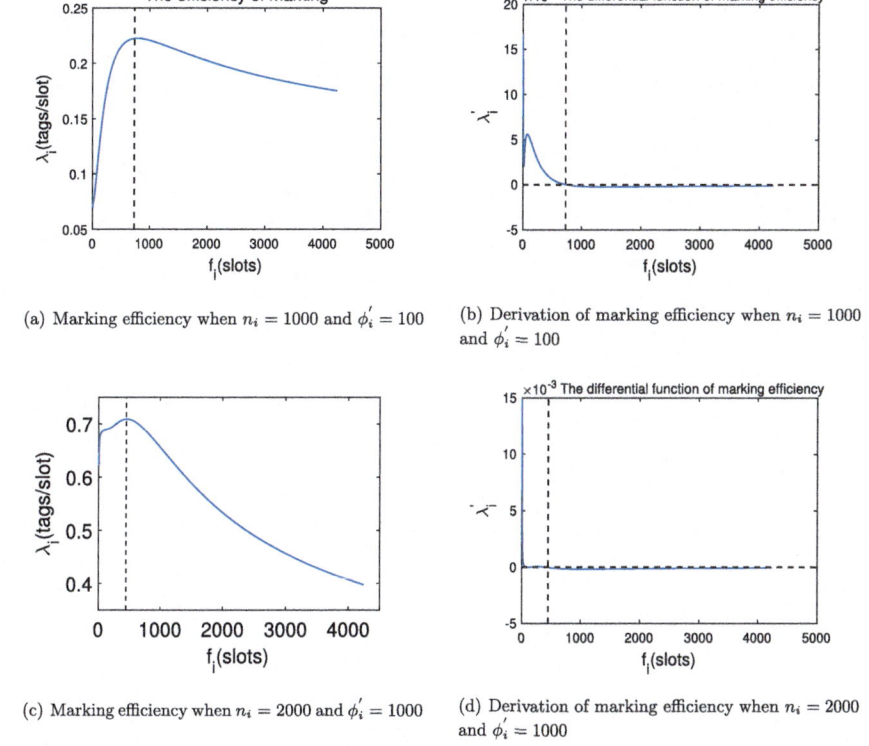

Fig. 4.2 An illustration of properties of marking phase λ_i vs. original vector size f_i

To understand the properties of λ_i, we depict its numerical results with $\frac{96}{T_{id}}$ omitted in Fig. 4.2 under diverse n_i and ϕ'_i. It can be observed that there exists an optimum f_i maximizing λ_i, which matches with the analysis stated in Theorem 6.

(2) Optimum parameters for the detection phase: The execution time in this phase is mainly spent on the composite vector transmission and the tags' responses. It is written as

$$T_d = \frac{f_d \lceil \log_2(l+1) \rceil}{96} T_{id} + f_d T_{short}. \qquad (4.25)$$

Our goal is to minimize f with the constraint of the detection reliability requirement. We first derive the detection probability of our approach. Let n_A define the number of the representative tags marked in the first phase, and p_2 be the sampling ratio in the second phase. Then the probability $P_j(p_2)$ that j marked representative tags are sampled in the detection could be expressed as

$$P_j(p_2) = \binom{n_A}{j} p_2^j (1 - p_2)^{n_A - j}. \tag{4.26}$$

We then recursively derive the probability that an arbitrary slot is singleton after l mappings given a j:

$$Ps_l = Ps_{l-1} + (1 - Ps_{l-1}) \binom{j - r_{l-1}}{1} \left(\frac{1}{f_d}\right) \left(1 - \frac{1}{f_d}\right)^{j - r_{l-1} - 1}$$

$$r_l = \lfloor f_d Ps_l \rfloor. \tag{4.27}$$

Thus, the probability that an arbitrary slot is singleton in our protocol after l mapping is

$$P_l = \sum_{j=0}^{n_A} P_j(p_2) Ps_l. \tag{4.28}$$

Since an arbitrary tag is mapped to a singleton slot with the probability of $\frac{f_d P_l}{n_A}$, the missing event detection probability in the advanced protocol can be approximately derived as

$$P_d = 1 - \left(1 - p_1 + p_1 \left(1 - \frac{f_d P_l(p_2)}{n_A}\right)\right)^{M_a} = 1 - \left(1 - \frac{f_d P_l(p_2)}{|\mathcal{G}_A|}\right)^{M_a}. \tag{4.29}$$

Note that M_a is a given threshold. Consequently, we should pick f_d and p_2 so that $P_d \geq \alpha$.

To this end, we could fix the value of p_2 and $P_d(p_2, f_d)$ is degraded into a function of f_d. Our goal is then turned to minimize f_d with $P_d(p_2, f_d) \geq \alpha$. After getting the optimum f_d for a given p_2, we start to introduce how to select p_2. When the sampling probability is too small to satisfy $P_d(p_2, f_d) \geq \alpha$, we cannot find suitable f_d. Hence we could set an upper bound for f_d. If $P_d(p_2, f_d) < \alpha$ when f_d is greater than the upper bound, we should increase the sampling probability p_2 to do another search. Finally, we could find the minimum sampling probability p_{min} that just satisfies the requirement. Then we will search the minimum f_d in $p_{min} \leq p_2 \leq 1$.

Now, we will discuss the influence of the value of multiple mapping. Fixing f_d while increasing l, we observe that the improvement shrinks rapidly from $l = 7$ to 15, since a bigger l would increase execution time according to (4.25). Therefore, we can search for the optimal value of l.

4.6 Performance Evaluation

In this section, we evaluate the performance of the proposed basic and advanced protocols in terms of detection probability and execution time in multi-tagged RFID systems. The timing parameters in the simulation follow the EPC-global Gen2 standard. Specifically, any two consecutive communications between the reader and tags are separated by a blank interval

4.6 Performance Evaluation

lasting for 266.4 μs. The transmission rate is 40.97kb/s when a response slot T_{short} is 290.81 μs and a 96-bit slot T_{id} is 2609.76 μs, which includes a blank interval. The parameters like the filter and vector size are set according to the theoretical analysis. In the simulation, we verify the effectiveness of the two protocols in addressing the missing event detection problem, where the results are obtained from 1000 independent runs. We also investigate the impacts of system scale and the number of tags on one object on their performance.

Performance Verification: We here verify the effectiveness and the efficiency of the proposed protocols under three scenarios. In the simulation, the threshold of missing objects is set to $M_a = 2$, and the required detection reliability varies from $\alpha = 95\%$, to $\alpha = 99\%$ and to $\alpha = 99.9\%$ in the first two scenarios and is fixed to $\alpha = 95\%$ in the third scenario. (1) In the first scenario, there exist 10 tags on each object and the number of overall tags varies from 1000 to 5000. The simulation results of detection probability and execution time are depicted in Figs. 4.3 and 4.4. The results show that both the bloom filter-based basic protocol and the compressive filter-based advanced one can meet the detection reliability requirement and they spend more time detecting a missing event as the number of overall tags increases. This can be interpreted as follows: As the number of objects increases, there are more representative tags that need to be marked and detected, leading to a longer execution time.

We can also observe that the advanced protocol needs significantly less time to detect missing events than the basic one under the same required detection reliability. As shown in Fig. 4.4c, when the number of total tags is 5000, the execution time of the basic protocol is 1.24s while the advanced protocol spends 0.38s which is 3x faster than the basic protocol. (2) In the second scenario, we study how the number of tags on one object influences detection probability and execution time. To this end, we set the total number of tags in the system to 1000 and vary the number of tags in each object A from 2 to 10. From the results recorded in Figs. 4.5 and 4.6, we can draw similar conclusions to those in the first scenario that both protocols can complete the detection task with the required reliability satisfied, and the advanced protocol is more time-efficient. In addition, the performance gain in terms of the execution time of the advanced protocol is at least 2x, and reaches 4x when the required detection reliability is 99.9% and there are two tags on each object, as shown in Fig. 4.6c. (3) In the third scenario, we focus on the time efficiency of the two protocols in large-scale systems, which is one of the most important metrics in RFID-enabled applications. The experiment consists of two cases: The number of tags on each object is fixed to 10 and the number of total tags varies from 5000 to 30000 in the first case; in contrast, we set the number of total tags to 30000 but change the number of tags on each object from 2 to 10 in the second case.

Figure 4.7a illustrates the impact of system scale on the execution time in the first case. We can observe that the two protocols experience longer execution times as the system scales up. However, the advanced protocol performs better. Figure 4.7b records the simulation results

Fig. 4.3 The achieved detection probability with the number of total tags varied from 1000 to 5000 when the threshold of missing objects is $M_a = 2$ and the required detection probability is **a** $\alpha = 95\%$, **b** $\alpha = 99\%$ and **c** $\alpha = 99.9\%$

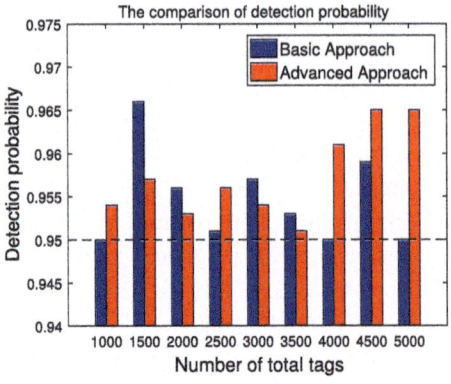

(a) $\alpha = 95\%$

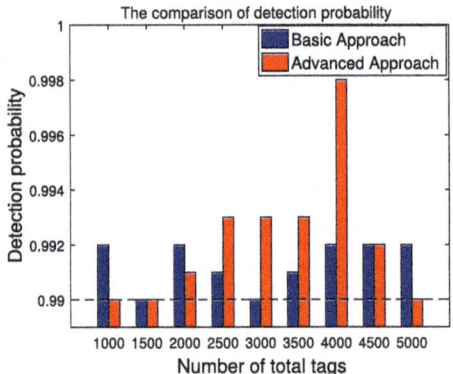

(b) $\alpha = 99\%$

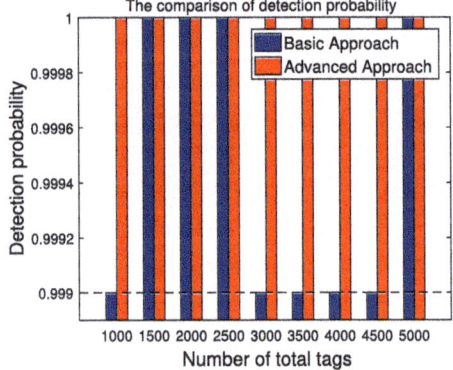

(c) $\alpha = 99.9\%$

4.6 Performance Evaluation

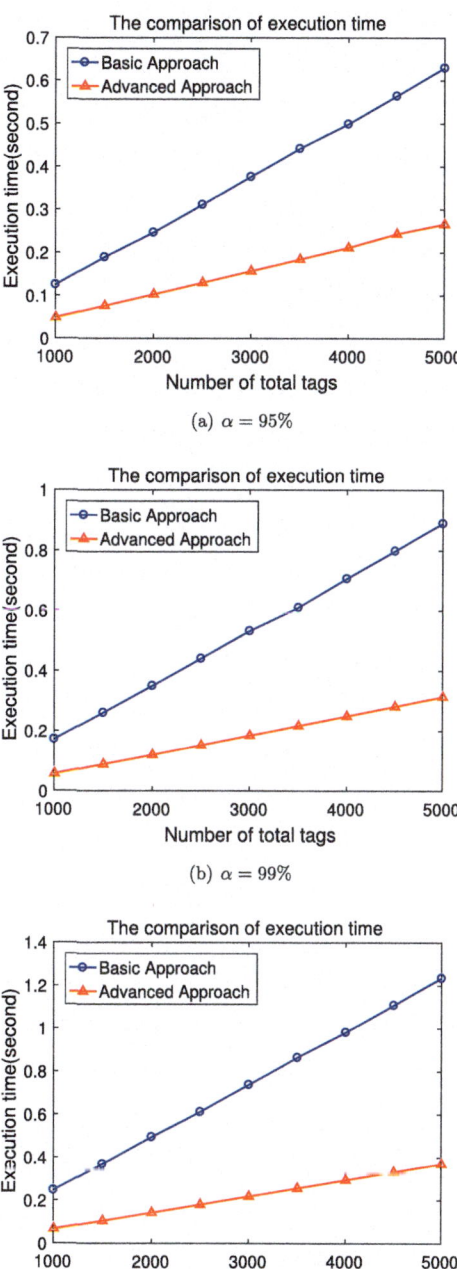

Fig. 4.4 The execution time with the number of total tags varied from 1000 to 5000 when the threshold of missing objects is $M_a = 2$ and the required detection reliability is **a** $\alpha = 95\%$, **b** $\alpha = 99\%$ and **c** $\alpha = 99.9\%$

Fig. 4.5 The detection probability with the number of tags on each object varied from 2 to 10 when the number of total tags is set to 1000, the number of missing objects is $M_a = 2$ and the required detection probability is **a** $\alpha = 95\%$, **b** $\alpha = 99\%$ and **c** $\alpha = 99.9\%$

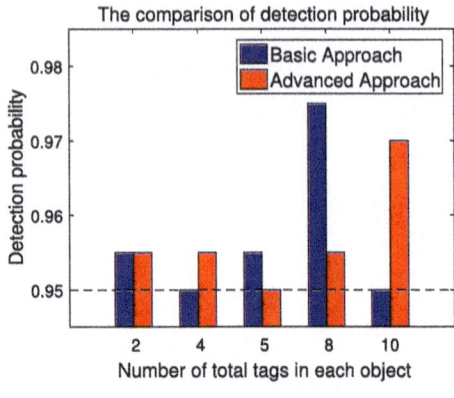

(a) $\alpha = 95\%$

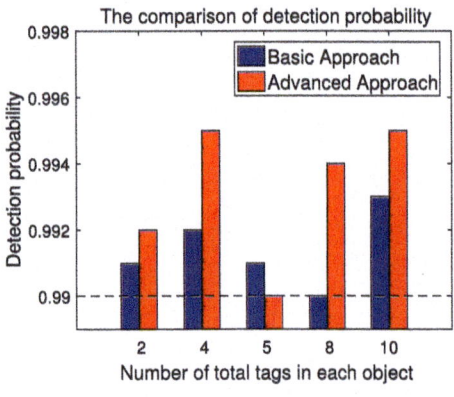

(b) $\alpha = 99\%$

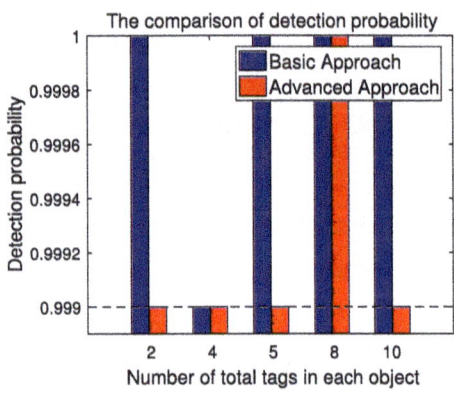

(c) $\alpha = 99.9\%$

4.6 Performance Evaluation

Fig. 4.6 The execution time with the number of tags on each object varied from 2 to 10 when the number of total tags is set to 1000, the number of missing objects is $M_a = 2$ and the required detection probability is **a** $\alpha = 95\%$, **b** $\alpha = 99\%$ and **c** $\alpha = 99.9\%$

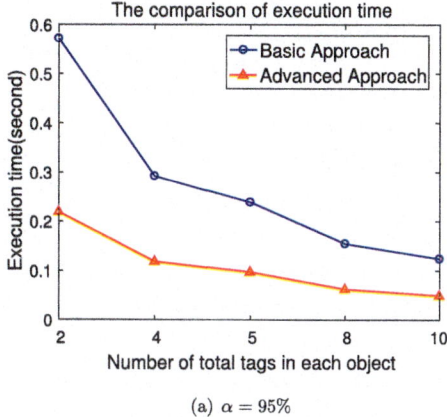

(a) $\alpha = 95\%$

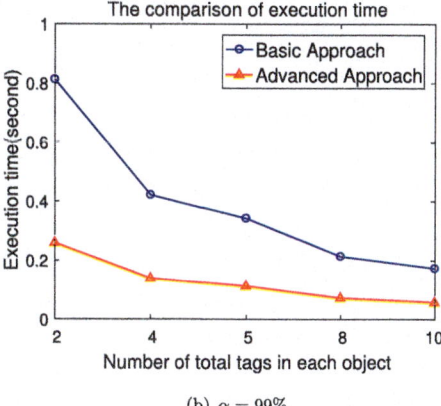

(b) $\alpha = 99\%$

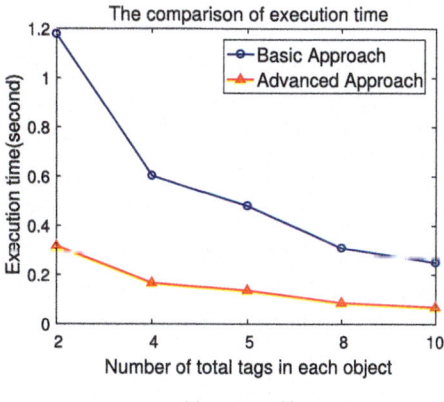

(c) $\alpha = 99.9\%$

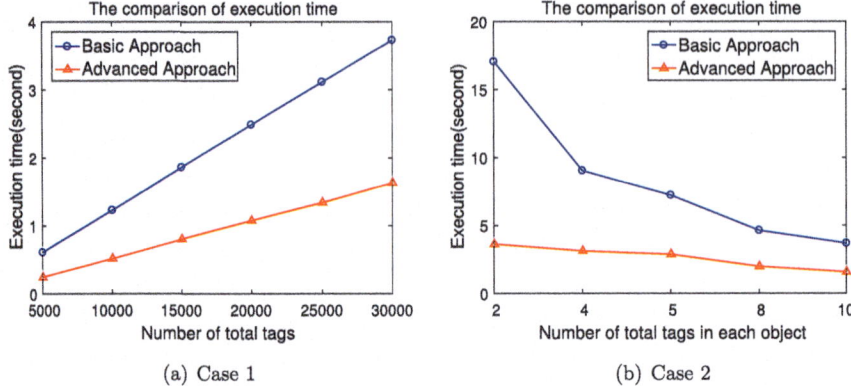

Fig. 4.7 a The execution time with the number of tags varied from 5000 to 30000 when each object is attached by 10 tags and the required detection probability is $\alpha = 95\%$. **b** The execution time with the number of tags on each object varied from 2 to 10 when the number of total tags is 30000, required detection probability is $\alpha = 95\%$

in the second case, which also confirms the superiority of the advanced protocol to the basic one. Moreover, It can be observed from the two figures that the advanced protocol achieves at least 2x performance gain.

4.7 Conclusion

This chapter has addressed a variation on the missing event detection problem arising from multi-tagged RFID systems where each object is tagged by multiple tags. Application of prior works to the new problem suffers low time efficiency due to repeated checks of one object. To overcome this drawback, we have provided two solutions, namely the basic protocol and the advanced protocol. The former uses the Bloom filter to ask a subset of tags in the system to report their presence. The latter exploits the knowability of each tag mapping and sparsity of slots mapped only by tag(s) of the chosen subset to build a compact compressive filter and a composite vector from multiple mappings of each tag. We have also derived the optimum parameters used in the protocols and conducted extensive simulations. The results confirm the effectiveness of the protocols and the superiority of the advanced protocol in terms of time efficiency under required detection reliability.

References

1. C. C. Tan, B. Sheng, Q. Li, How to monitor for missing rfid tags, in *IEEE ICDCS* (2008), pp. 295–302
2. W. Luo, S. Chen, T. Li, Y. Qiao, Probabilistic missing-tag detection and energy-time tradeoff in large-scale rfid systems, in *ACM MobiHoc* (2012), pp. 95–104
3. W. Luo, S. Chen, Y. Qiao, T. Li, Missing-tag detection and energy-time tradeoff in large-scale rfid systems with unreliable channels. IEEE/ACM ToN **22**(4), 1079–1091 (2014)
4. M. Shahzad, A. X. Liu, Expecting the unexpected: Fast and reliable detection of missing rfid tags in the wild, in *IEEE INFOCOM* (2015), pp. 1939–1947
5. J. Yu, L. Chen, R. Zhang, K. Wang, Finding needles in a haystack: Missing tag detection in large rfid systems. IEEE TCOM **65**(5), 2036–2047 (2017)
6. J. Yu, L. Chen, R. Zhang, K. Wang, On missing tag detection in multiple-group multiple-region rfid systems. IEEE TMC **16**(5), 1371–1381 (2016)
7. J. Yu, W. Gong, J. Liu, L. Chen, K. Wang, R. Zhang, Missing tag identification in cots rfid systems: Bridging the gap between theory and practice, *IEEE TMC* (2018)
8. T. Li, S. Chen, Y. Ling, Identifying the missing tags in a large rfid system, in *Proceedings of the eleventh ACM international symposium on Mobile ad hoc networking and computing* (ACM, 2010), pp. 1–10
9. R. Zhang, Y. Liu, Y. Zhang, J. Sun, Fast identification of the missing tags in a large rfid system, in *2011 8th Annual IEEE Communications Society Conference on Sensor, Mesh and Ad Hoc Communications and Networks* (IEEE, 2011), pp. 278–286
10. X. Liu, K. Li, G. Min, Y. Shen, A.X. Liu, W. Qu, Completely pinpointing the missing rfid tags in a time-efficient way. IEEE ToC **64**(1), 87–96 (2015)
11. J. Yu, W. Gong, J. Liu, L. Chen, K. Wang, On efficient tree-based tag search in large-scale rfid systems. IEEE/ACM ToN **27**(1), 42–55 (2019)
12. L. Bolotnyy, G. Robins, Multi-tag rfid systems. Int. J. Internet Protocol Technol. **2**(3), 218–231 (2007)
13. D. Hochhalter, D. Bigelow, N. J. Witchey, C. Milam, Rfid-based rack inventory management systems, 2018. US Patent App. 15/725, 638
14. J. Yu, J. Liu, R. Zhang, L. Chen, W. Gong, S. Zhang, Multi-seed group labeling in rfid systems, in *IEEE TMC* (2019)
15. W. Gong, J. Liu, Z. Yang, Fast and reliable unknown tag detection in large-scale RFID systems, in *ACM MobiHoc* (2016), pp. 141–150

Fast and Reliable Access Protocol for Multi-Tagged RFID Systems

This chapter designs a method for detecting missing multi-tagged events detection from the perspective of hashing seed. Our key idea is to search appropriate seeds such that the reader only needs to probe a subset of tags, each selected from different items rather than the entire tag set for the missing item detection. By employing the computation-communication trade-offs, we develop two protocols named M^2ID and M^2ID+, where the latter protocol improves time efficiency by classifying tags before segmentation compared to the former. With the derived optimum parameters, our protocols can achieve up to 4x performance gain in terms of time efficiency compared to state-of-the-art solutions.

Chapter roadmap: The rest of this chapter is organized as follows. Section 5.1 introduces the motivation of joint computation and communication on the multi-tagged detection RFID systems. In Sect. 5.2, we enumerate the missing event detection in RFID-based applications and show the time inefficiency in the multi-tagged items in RFID systems. Then, a suit of the segmentation and the seed searching method and an advanced method to improve time efficiency are proposed in Sects. 5.4 and 5.5, respectively. In Sect. 5.6, we evaluate the reliability and time efficiency of the proposed protocols. Finally, we conclude this chapter in Sect. 5.7.

5.1 Introduction

This chapter also concentrates on a particular missing item detection scenario arising from multi-tagged RFID systems. As mentioned in Chap. 4, prior works [1–10], cannot be applied in the missing multi-tagged item detection since here these tag-level algorithms would be time-inefficient, repeating checking the present item. Therefore, the Chap. 4 proposes the missing items detection based on the filter construction, which is filtering and detecting a subset of the entire tags to detect missing items. The simulation results illustrate the time efficiency of the proposed methods.

We study the missing multi-tagged item detection problem from the aspect of computation and communication, and develop two protocols. The first protocol named M^2ID provides a suit of the segmentation and the seed searching methods. M^2ID divides all tags into multiple tag segments to reduce the computation time cost of the seed searching. It makes each segment contain the same number of the target tags with the unique hashing values falling into a specific interval. This would reduce the communication time cost by avoiding broadcasting unique hashing values to the tags in sequence and avoiding receiving time-inefficient responses since these unique values imply the positions of the target tags' response slots are randomly distributed. On the top of M^2ID, we propose M^2ID+ to further improve the time efficiency. M^2ID+ works in a 'Sampling-Classification-Segmentation' pattern, which can reduce the reader-tag communication cost at the price of extra less computation cost and thus the overall time cost. The main contributions of this chapter are articulated as follows.

- We formulate the largely unaddressed missing multi-tagged item detection problem in RFID systems and provide solutions from the perspective of the computation-communication trade-offs.
- We present two concrete protocols namely M^2ID and M^2ID+. M^2ID constructs a framework integrating the seed searching and the reader-tag communication, enabling the computation-communication trade-offs. M^2ID+ exploiting the time difference between the computation and the communication reduces the communication cost at the expense of affordable extra computational cost, improving the time efficiency.
- We optimize the protocol performance with the optimum parameters derived. The analytical results also reveal the relationship between the frame size and the probability of finding a proper seed, indicating the computation-communication trade-offs.
- We conduct extensive simulations to evaluate the performance of the proposed protocols. The results show that M^2ID and M^2ID+ achieve performance gain of 2x and 4x over the state-of-the-art one [11] in terms of time efficiency, respectively.

We would like to emphasize that we provide not only efficient solutions to the missing multi-tagged item detection problem but also a new methodology embracing the computation-communication trade-offs, which benefits solving other protocol design problems in RFID systems.

5.2 Related Work

The works of missing item detection are separated into two categories: probabilistic protocols [1–7] and deterministic protocols [8–10].

Probabilistic protocols detect a missing item event with a predefined probability. Tan et al. initiate the study of probabilistic detection and propose a solution called Trusted Reader Protocol (TRP) in [1]. TRP detects missing item events by comparing the pre-computed slots with those picked by the tags attached to items. If an expected singleton slot turns out to be empty, then the missing item event is detected. Luo [2] and [3] employ multiple seeds to increase the probability of the singleton slot, which reduces the useless empty and collision slots and thus achieves better performance. RUN [4] and BMTD [5] address the influence of the unknown tags. Yu et al. design a suit of detection protocols for multi-categories and multi-region RFID systems and study how to detect missing items by using COTS RFID devices [7].

Deterministic protocols, on the other hand, are able to exactly identify which items are absent. Li et al. develop a series of deterministic protocols in [8] to reduce the radio collision by reconciling collision slots and finally iron out a bit-level tag identification method by iteratively deactivating the tags of which the presence has been verified, hence affirming the presence of items. Subsequently, Zhang et al. propose [9] to identify tag responses in all rounds and observe the change among the corresponding bits among all bitmaps to determine the present and the absent tags for identifying the presence of items. However, how to configure the protocol parameters is not theoretically analyzed. More recently, Liu et al. [10] enhanced the work by reconciling both 2-collision and 3-collision slots and filtering the empty and unreconcilable collision slots to improve time efficiency.

With the presence of multi-tagged items in RFID systems, the prior works show their weakness in terms of time efficiency. The key to addressing the multi-tagged missing item detection problem is to probe a subset of the tags for the detection, avoiding repeated checks of the present items. However, very limited works have studied the problem from this perspective. The most related work [11] utilizes a bloom filter to solve the tag identification problem for multi-tagged RFID systems. Yet, the false positives of the bloom filter and the low ratio of the singleton slots (no more than 36.7%) make it time-inefficient for the missing item detection.

5.3 System Model and Problem Formulation

In this section, we will introduce the system model used in our chapter and formulate the problem of detecting the missing multi-tagged items in an RFID system.

5.3.1 System Model

We consider an RFID system consisting of one reader[1] and a large number of tags, where each physical item is attached by multiple tags. The reader is connected with a back-end server which has a powerful computational capability. For the purpose of simplification, we treat the reader and the server as an entity and just call it the reader. Moreover, each tag has a unique ID and performs computations such as hashing function. All tags' IDs in the system are recorded by the reader.

The communication between the reader and the tags follows the rule of 'Listen-before-talk' [13, 14]. In the detection, for example, the reader broadcasts commands and parameters including the frame size f and a seed s at first. Then, each tag uses its ID and the received seed s to generate one pseudo-random value via hash function as $(H(ID, s) \mod f)$, and executes the next step according to the received commands (i.e., compare, response, or wait for next commands).

The downlink (i.e., reader-to-tags) transmission is continuous. The uplink (i.e., tags-to-reader), on the other hand, contains a blank slot between any two tags' 1-bit responses [13]. For simplicity, we denote T_d and T_{tag} as the time duration of 1-bit broadcasting slot and 1-bit response slot, respectively. Consider an arbitrate response slot, it may experience three states. When no tags respond in this slot, it is an empty slot; when a single tag responds, it is a singleton slot; when multiple tags respond, it is a collision slot. The latter two states are also regarded as non-empty slots. Considering the unstable channel, there exist error transmissions. We assume that the downlink works in an error-free channel since the reader supported by the external power source can increase the transmission power. In contrast, the tag cannot support much power to counter the interference. Therefore, we assume that the error occurs in the uplink and the manifestation of the error transmission is bit inversion. The '1' inverted to '0' brings the false positives and the '0' inverted to '1' induces the false negatives which will cause the practical damage.

5.3.2 Problem Formulation

In this chapter, we are interested in detecting the missing multi-tagged items in an RFID system where n tags monitor g items each attached by the multiple tags, i.e., $g < n$. Considering the instability of the uplink, we define m_a as the number of missing items and P_d as the probability that the reader succeeds in detecting the missing item event without the false alarm. We formulate the multi-tagged missing item detection problem as follows: The missing multi-tagged item detection problem is to devise an algorithm of minimum execu-

---

[1] For multiple readers, we can treat them as a single virtual reader as in [6, 12]. Specifically, the back-end server searches all proper seeds and corresponding hashing values and sends them to all readers such that readers broadcast these parameters. Consequently, the back-end server can synchronize the readers and we can logically consider them as a whole.

5.3 System Model and Problem Formulation

tion time to detect missing item event with $P_d \geq \alpha$ under the condition of $m_a \geq M_a$ in the unstable uplink. The α is the required correct detection probability among all detections, and the M_a is a predefined detection threshold meaning the tolerance to the minimum number of the missing items. Note that the problem is degraded to deterministically identify missing items when $\alpha = 1$ and $M_a = 1$. The proposed protocols in this chapter can also achieve deterministic missing item identification.

We would like to emphasize the key difference between the missing multi-tagged item detection problem and the prior missing single-tagged item detection problem: The successful response of one tag on a multi-tagged item indicates the presence of the item, it is thus feasible to query one tag for checking the state of an item. In contrast, if we use the prior algorithms for the missing single-tagged item detection problem, all tags on the item would respond to the interrogation, resulting in severe interference and thus considerably degrading time efficiency. For example, there are 10,000 items being monitored where each item is attached by 3 tags, the prior works have to probe 30,000 tags, which sharply increases the time cost. Instead, we only query 10,000 tags by picking one tag from each item. Table 5.1 summarizes the main notations used in this chapter. The difference makes our work different from the prior ones and more challenging.

5.3.3 Design Rational

The response of a tag means the presence of the item, the reader has no need to query the other tags on this item. Meanwhile, the absence of one tag indicates the potential missing item. Therefore, it is feasible to probe one of the tags on an item instead of all for the missing item event detection. If the probed tag is absent, we would further poll the left tags on the item, and a missing item would be found if all of them are absent. Considering the number of missing items is usually small, the idea above can improve time efficiency significantly.

In this chapter, we randomly select one tag from each item, referred to as a representative tag. These g tags constitute the representative tag set defined as $\mathcal{G}_A = \{RT_1, RT_2, ..., RT_g\}$ where the RT_k is a tag on the item k for $1 \leq k \leq g$. The set of the remaining tags named pending tags is denoted by $\mathcal{G}_B$. We are interested in interrogating the representative tags to detect potential missing item events. Unfortunately, the pending tags in $\mathcal{G}_B$ would cause severe interference to the representative tag detection. Hence, an efficient scheme should be able to eliminate this negative impact.

We distinguish the representative tags and the pending tags via their hashing values by selecting such a seed that there exists no common hash value mapped by the tags in $\mathcal{G}_A$ and $\mathcal{G}_B$. Furthermore, we prefer a proper seed that makes each tag in $\mathcal{G}_A$ map to a unique hash value and respond in a singleton slot. Consequently, the reader only needs to broadcast the proper seed and corresponding unique hashing values. Each tag learns whether it should respond after comparing its hashing value with the received hashing value, and the

Table 5.1 Main parameter notation

Symbols	Description
n	The number of the tags in our system
g	The number of the items in our detection region
f_{sample}	The frame size of the hashing for sampling
s_{sample}	The seed used for sampling
f_{sg}	The frame size of doing hash in the segmentation
s_{sg}	The seed used in the segmentation
b_{is}	The lower bound of the i-th segment
b_{ie}	The upper bound of the i-th segment
p_{s_i}	The probability of searching a proper seed in the i-th segment
N_{s_i}	The round of finding out a proper seed in the i-th segment
g_d	The fixed number of representative tags in each segment
N_c	The round of the CPU clock doing once hash
T_{CPU}	The period of the CPU clock
$L(\cdot)$	The operation of $\log_2(\max\{\cdot\})$
f_{seg_u}	The frame size for segmentation in category u
f_{s_u}	The identical frame size for seed searching for all segments
n_{q_u}	The number of tags in the q-th segment in category u
T_d	The period of broadcasting a bit
T_{tag}	The period of the tag's 1-bit response in a slot

response slot of the representative tag is determined by the hashing value. After receiving the representative tags' responses, the reader can find the potential missing items.

Yet, it is impractical to search a proper seed for all tags in a large-scale system of extremely high computational complexity, it is necessary to design a strategy that can lessen the number of tags simultaneously involved in the seed searching. On the other hand, these unique hashing values indicating the response slot positions of the representative tags are randomly distributed. Hence, we have to broadcast each of these hashing values and the response slots might involve empty slots, leading to low efficiency. To address the obstacles, we induct these unique hashing values to a special range so that we only need to broadcast the boundary values of this range once and all response slots are singleton, which retrenches the reader-tags communications. This makes the communication cost from broadcasting hashing values many times determined by the number of the representative tags to broadcasting only twice, meanwhile, from involving empty slots to reaching 100% utilization in the response frame.

Moreover, motivated by the difference between the computation time of the reader and the transmission rate among the reader-tag communications (i.e., the clock period of the

CPU in the reader is about 0.3ns and a time slot in the communication is over $10\mu s$), we could trade-off the computation cost and the communication cost to further minimize the overall execution time. Following the design rational above, we construct M^2ID and the improved M^2ID+.

5.4 M^2ID: Missing Multi-Tagged Item Detection Protocol

In this section, we first introduce the Missing Multi-tagged Item Detection Protocol (M^2ID) and analyze how to tune the parameters for performance optimization.

5.4.1 Motivation

The seed and a tag's ID determine the hashing value of the tag in RFID systems, so we can find a proper seed that makes each of g representative tags have a unique hashing value. The reader then broadcasts the seed and the corresponding hashing values informing the tags whether/when they should respond. We denote the j-th representative tag's hashing value under the proper seed s and the frame size f by $h_j = (H(ID_j, s) \mod f) + 1$, meaning its response slot is also h_j. To avoid sequentially broadcasting these unique hashing values, we make them all designated in a range of $[1, g] \subseteq [1, f]$ that we only need to broadcast the parameter g once at the beginning of the protocol. This makes the communication cost from broadcasting every unique hashing value to broadcasting only one value.

We take an example to illustrate this. As shown in Fig. 5.1, there exists 2 items and each is attached with 3 tags (i.e., $g = 2$ for 2 representative tags). We pick one tag from each item as a representative tag (i.e., tag 1,4). Originally, we select a proper seed s' so that all representative tags' hashing values are unique and their corresponding hashing values represent the slot position in the response frame. For example, the hashing value of the tag 1 is 2 and it should respond in the 2-th slot in the response frame, the tag 4's hashing value is 5 and it should respond in the 5-th slot in the response frame (c.f. Fig. 5.1a). Hence, we have to broadcast these hashing values (i.e., 2, 5) to tags and the response frame involves empty slots, leading to low efficiency. In our design, as shown in Fig. 5.1b, our required seed s makes all unique hashing values fall in the range of $[1, 2]$, while the pending tags' hashing values are out of this range. Finally, tag 1,4 will respond in the response slots 1,2, respectively. Thus, we only need to broadcast hashing value 2 and the utilization of the response frame reaches 100%, which retrenches communication time cost.

Finding such a seed is effective for the missing item detection, but the time cost would soar if we search for all tags in the system. Specifically, the probability of seeking out the proper seed can be expressed as follows.

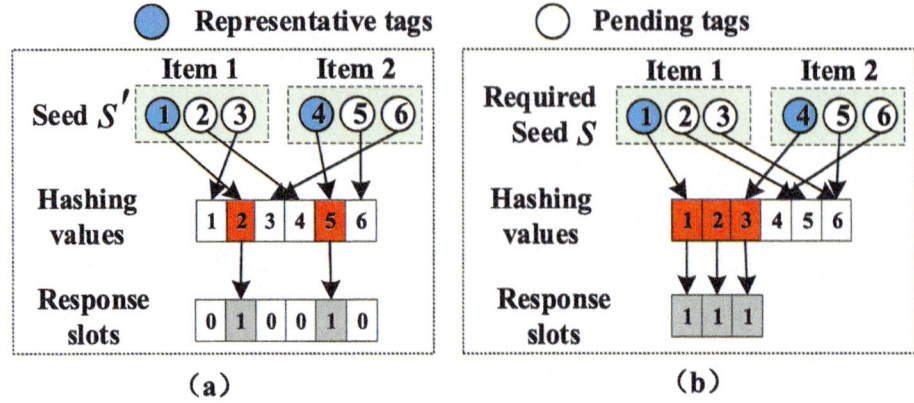

Fig. 5.1 The example of required seed s when $n = 6$, $g = 2$, $f = 6$

$$p_{ft} = \frac{g!}{g^g}\left(\frac{g}{f}\right)^g \left(1-\frac{g}{f}\right)^{n-g}. \tag{5.1}$$

It can be proven that p_{ft} is maximum when $f = n$ (c.f. Sect. 5.4.4). We next show that the time cost is unaffordable even for a small-scale system. For example, when $n = 100$ and $g = 50$, we have $p_{ft} = 2.701 \times 10^{-51}$. That is to say, we need $N_s \times T_{cpu} \approx 3.702 \times 10^{38}$s $\approx 1.174 \times 10^{31}$ years on average with 1,000 GHz CPU to find the seed for 100 tags. Therefore, a scheme of lower computation complexity is called for to achieve time-efficient detection.

In this chapter, we follow the principle of 'Divide and Conquer'. Take the system of $n = 100$ and $g = 50$ as an example again. We first divide them into 10 segments and each segment consists of $n_i = 10$ tags including $g_i = 5$ representative tags. Recall (5.1), it only takes $5.333\mu s$ to finding the proper seed in a segment when the reader works on a 5GHz CPU. The total searching time for all segments is $53\mu s$, which is significantly less than the unsegmented one above. Aggressively, if we divide all tags into 50 segments where $n_i = 2$, $g_i = 1$ on average, the total searching time decreases to 40ns. Therefore, the segmentation is an effective method to decrease the time cost in the seed searching.

5.4.2 Segmentation

Segmentation can reduce the computational complexity, but directly segmenting the tags randomly is undesirable. Recall that the tags conduct hashing operations with a seed and the frame size, and are segmented by their hashing values and a given segment size. The random segmentation would unbalance the number of the tags falling into the segments, degrading the performance gain of the segmentation. On the other hand, we have to spend

5.4 M²ID: Missing Multi-Tagged Item Detection Protocol

extra time on telling tags the range of unique hashing values in each segment due to the different number of the representative tags in each segment.

In this chapter, we propose a segmentation method of approximately uniformly segmenting the set of tags. By uniform segmentation, each segment contains an identical number of the representative tags and a similar number of pending tags. Hence, the hashing value range of each segment might be different. More specifically, this method operates as follows. First, the reader records all tags' IDs, we thus simulate at the reader that all tags calculate their hashing values under a seed s_{sg} and frame size f_{sg}. The setting of frame size f_{sg} will be discussed in Sect. 5.4.4.

Second, the reader determines the hashing value range of each segment guaranteeing the identical number of the representative tags in each segment. Each segment's hashing value range can be expressed via boundary values. The lower bound and upper bound values can be set as follows. We denote g_i as the number of the representative tags in the i-th segment for $i = 1, 2, \ldots$. For the first segment, the reader sets the lower bound as $b_{t1} = 1$ and then seeks out the largest value of the upper bound b_{e1}. The required b_{e1} should satisfy $g_1 = \sum_{k=b_{t1}}^{b_{e1}} \Phi(k) \leq g_d$, where g_1 is the number of the representative tags whose hashing values are in the range of $[b_{t1}, b_{e1}]$, $\Phi(k)$ is the number of the representative tags whose hashing values equal to k, and g_d is the required number of the representative tags in each segment. Once b_{e1} set, the boundary values of the first segment are set as $[b_{t1}, b_{e1}]$. Thus, the tags should be in 1-st segment when their hashing values are into the range of $[b_{t1}, b_{e1}]$. For the second segment, we set the lower bound as $b_{t2} = b_{e1} + 1$ and find the largest value of b_{e2} as the first segment does. After repeating the above processes, we eventually set each segment's boundary values so that each segment contains no more than g_d representative tags.

5.4.3 Protocol Description

Our proposed Multi-tagged Missing Item Detection Protocol M²ID can be described as follows.

We start at the view of the reader. (1) In the segmentation, the reader picks up an arbitrary seed s_{sg} from the seed pool and encodes each tag via its hashing value which is in a range of $[1, f_{sg}]$, meanwhile, confirms the boundary values of each segment. Then, the reader compares each tag's hashing value with the segment boundary to decide each tag's affiliation. (2) In the seed searching, the reader searches a proper seed under the corresponding optimum frame size to separate the representative tags and the pending tags in each segment. The position of a tag's response is determined by the serial number of the segment and its hashing value. In the i-th segment, the representative tags' hashing values are unique and in a range of $[1, g_d]$ with the proper seed s_i and the optimum frame size f_i. In case that we cannot always seek out the proper seed for all representative tags mapping to $[1, g_d]$ since the limited scale of the seed pool, we prefer the sub-optimum seed that makes the representative tags map to $[1, g_d]$ as many as possible. **(3)** In the reader-tag communications, the reader first broadcasts parameters including the seed s_{sg} and the frame size f_{sg} for the segmentation. Then, it broadcasts the boundary values $\{b_{ti}, b_{ei}\}$ of the i-th segment and its corresponding seed s_i and the frame size f_i. After finishing broadcasting, the reader interrogates the tags and listens to the tags' responses for detecting the missing item event.

At the tags end, each tag receives parameters from the reader and first calculates its hashing value for the segmentation with the seed s_{sg} and the frame size f_{sg}. Then, it compares its hashing values with the boundary values of segments. If its hashing value falls into $\{b_{ti}, b_{ei}\}$, it should be in the i-th segment. After that, it uses the received s_i and f_i to do another hashing. If the value falls into $[1, g_d]$, the tag will regard itself as a representative tag and then respond in the corresponding position of the response frame when interrogated. Otherwise, the tag will keep silent and wait for the reader's new command. Consequently, only the representative tags respond and all the response slots should be singleton.

Since the response slots are orchestrated to be mapped by one representative tag, the reader knowing all representative tags' mapping positions can detect the missing item even if there exists at least one empty slot. After finding any empty slot, the reader will poll the other tags on the item attached by this missing representative tag to affirm the item state.

By conducting the operations above across all segments, we can identify all representative tags. One of the challenges in M^{2}ID is how to tune parameters for the minimum execution time. We will address this in the following.

5.4.4 Parameter Optimization

In this part, our optimization is described based on the situation that all proper seeds are sought out in the seed pool. Our goal is to configure the frame size f_{sg} for the segmentation, the frame size f_i for the seed searching in the i-th segment, and the required number of the representative tags g_d in each segment.

(1) At the beginning, we first discuss the setting of the frame size f_{sg} in the segmentation. Based on the description in Sect. 5.4.2, the number of the representative tags in each segment is no more than g_d. Thus, the positions of the representative tags' responses are decided as follows:

$$RePos_j = g_d(i-1) + y_j, \quad (5.2)$$

where the y_j is the hashing value of the j-th representative tag in the i-th segment. In a segment, g_i representative tags' hashing values are unique and in the range of $[1, g_d]$. Sometimes, g_i may be smaller than g_d because of the coincident hashing values of the multiple representative tags. If we make $g_i = g_d$ in each segment, we would achieve 100% utilization of the response frame without empty slots. For example, each segment has two representative tags and we consider the i-th segment. The proper seed in this segment makes

5.4 M²ID: Missing Multi-Tagged Item Detection Protocol

the representative tags' hashing values $y_1 = 1$ and $y_2 = 2$. Thus, the tags will respond in the $(2(i-1)+1)$-th slot and the $(2(i-1)+2)$-th slot, respectively. The key to achieving $g_i = g_d$ is that each representative tag maps to a unique value without considering the pending tags' hashing values. The probability that each representative tag maps to a unique hashing value can be written as

$$p_g = \frac{\prod_{j=0}^{g-1}(f_{sg}-j)}{f_{sg}^g} \geq 66.7\%. \tag{5.3}$$

The expected round of the seed searching defined as N_g is $1/p_g$. Here, $p_g \geq 66.7\%$ means that an arbitrary seed makes the representative tags map to the unique values, i.e., $\mathbb{E}[N_g] = 1/p_g \leq 1.499 \approx 1$. The value of f_{sg} can thus be derived while the number of the representative tags in each segment is g_d.

(2) We next discuss the optimum frame size f_i in the seed searching. The probability of seeking out a proper seed for the i-th segment can be expressed as follows:

$$p_{s_i} = g_d! \left(\frac{1}{f_i}\right)^{g_d} \left(1 - \frac{g_d}{f_i}\right)^{n_i - g_d} \tag{5.4}$$

where n_i is the number of the tags in the i-th segment and f_i is the frame size for calculating the hashing value. We then should derive f_i to maximize p_{s_i}.

Theorem 1 *Given g_d and n_i in the i-th segment, the optimum size of f_i maximizing p_{s_i} should satisfy $f_i = n_i$.*

Proof The partial differential function of p_{s_i} by f_i can be expressed as

$$\frac{\partial p_{s_i}}{\partial f_i} = \frac{g_d g_d!}{f_i^{g_d+1}} \left(1 - \frac{g_d}{f_i}\right)^{n_i - g_d - 1} \left(\frac{n_i}{f_i} - 1\right). \tag{5.5}$$

When $\frac{\partial p_{s_i}}{\partial f_i} = 0$, we have two points such as $f_{i_1} = g_d$ and $f_{i_2} = n_i$, and $g_d \leq f_i$. According to the requirement of the proper seed, we set $f_i = n_i$. As shown in (5.5), it is positive when $f_i < n_i$ and negative when $f_i > n_i$. Hence, we can get the maximum p_{s_i} with $f_i = n_i$.

Theorem 1 indicates the optimum frame size f_i is determined by the number of the tags n_i in this segment.

(3) Now, we discuss the selection of g_d. The expected execution time of the segmentation, defined as T_{sg}, consists of the time spent on the seed searching and broadcasting the parameters including the seed value, the frame size for the segmentation, and each segment's boundary values. The cost of broadcasting g_d can be negligible compared with the cost of other parameters broadcasting. Thus, the cost of the segmentation and broadcasting of each boundary value for all segments can be expressed as

$$T_{sg} = T_{searching} + T_{each} = nN_g N_c T_{cpu} + (L(f_{sg}) + L(s_{sg})) T_d + \sum_{i=1}^{r}(L(b_{ti}) + L(b_{ei})) T_d, \quad (5.6)$$

where $N_g = 1$ holds following (5.3). We use $L(\cdot)$ to stand for $\log_2(max\{\cdot\})$ and the number of the segments r is equal to g/g_d. The operator of $L(\cdot)$ shows the length of data expressed by the binary sequence. As the boundary values of each segment should be less than f_{sg}, the length of the binary sequence expressing boundary values is twice of $L(f_{sg})$. Therefore, (5.6) can be rewritten as

$$T_{sg} = nN_c T_{cpu} + \left(L(s_{sg}) + \left(\frac{2g}{g_d} + 1\right) L(f_{sg}) \right) T_d. \quad (5.7)$$

We can observe that the execution time of the segmentation is decided by g_d when f_{sg} is derived by (5.3).

Then, we will discuss the execution time after each tag recognizes its belonged segment. As described in Sect. 5.4.3, the expected total execution time of the seed searching, broadcasting the corresponding optimum frame sizes and the seed values can be written as

$$T_s = \sum_{i=1}^{r} N_{s_i} n_i N_c T_{cpu} + (L(f_i) + L(s_i)) T_d, \quad (5.8)$$

where the N_{s_i} is the round of the seed searching and we have $N_{s_i} = 1/p_{s_i}$. As stated in the Theorem 1, we prefer p_{ms} to represent the maximum of p_{s_i} with the respect to f_i if we make $f_i = n_i$:

$$p_{ms}(g_d, n_i) = \frac{g_d!}{g_d^{g_d}} \left(\frac{g_d}{n_i}\right)^{g_d} \left(1 - \frac{g_d}{n_i}\right)^{n_i - g_d}. \quad (5.9)$$

Hence, the minimum expected round for searching the proper seed is $1/p_{ms}(g_d, n_i)$.

During the searching process, we initialize the seed value to '1' and increase by '1' in the next round if we do not find out the proper one. Therefore, the value of s_i is equal to $1/p_{ms}(g_d, n_i)$, and T_{ss} can be rewritten as

$$T_s = \sum_{i=1}^{r} \frac{n_i N_c T_{cpu}}{p_{ms}(g_d, n_i)} + \left(L(n_i) + L\left(\frac{1}{p_{ms}(g_d, n_i)}\right) \right) T_d. \quad (5.10)$$

In addition, the execution time of the response frame can be expressed as

$$T_r = \sum_{i=1}^{r} g_i T_{tag} = r g_d T_{tag} = g T_{tag}. \quad (5.11)$$

5.4 M²ID: Missing Multi-Tagged Item Detection Protocol

Recall (5.6), (5.10), and (5.11), the expected execution time of the M²ID, defined as the T_{whole}, is

$$T_{whole} = T_{sg} + T_s + T_r = nN_c T_{cpu} + gT_{tag}$$
$$+ \left(L(s_{sg}) + \left(\frac{2g}{g_d}+1\right) L(f_{sg})\right) T_d + \sum_{i=1}^{\frac{g}{g_d}} \frac{n_i N_c T_{cpu}}{p_{ms}(g_d, n_i)} + \left(L(n_i) + L\left(\frac{1}{p_{ms}(g_d, n_i)}\right)\right) T_d. \quad (5.12)$$

We observe that T_{whole} is determined by the latter part of T_{sg} (i.e., $2g/g_d L(f_{sg})T_d$) and the whole part of T_s. Considering f_i and f_{sg} are fixed so that T_d and T_s are determined by g_d, we thus only optimize these two part with g_d where we approximate n_i with the expected number of the tags $\mathbb{E}[n_i] = ng_d/g$. However, using the expected value of n_i may be inaccurate for $L(n_i)$ that depends on the maximum n_i. To solve this problem, we set an upper bound for the maximum n_i. Based on extensive numerical analysis, we fix max $\{n_i\} = 3n/g$. Intuitively, we extract the part related with g_d from (5.12), the expression is written as follows:

$$T_{sim} = \frac{g}{g_d}\left(\frac{ng_d}{g} \frac{N_c T_{cpu}}{p_{ms}(g_d, \mathbb{E}[n_i])}\right) + \frac{2g}{g_d} \log_2(f_{sg}) T_d$$
$$+ \frac{g}{g_d}\left(\log_2\left(\frac{3ng_d}{g}\right) + \log_2\left(\frac{1}{p_{ms}(g_d, \max\{n_i\})}\right)\right) T_d. \quad (5.13)$$

Now, the problem is converted to find out the proper value of g_d to minimize T_{sim}. Generally, we are going to directly conduct the differential function of T_{sim} to calculate the extreme point where the proper g_d can be found. Yet, the differential function is too complex to derive the closed form of g_d. A feasible way is to find such an upper bound that the values of g_d over this bound would make T_{sim} increasing.

Conducting algebraic operations, we can observe that the first part of T_{sim} is of the order of the magnitude $\Theta((\frac{ne}{g})^{g_d})$ while the sum of the other parts is in $\Theta(\frac{1}{g_d})$. Consequently, we can find the upper bound for g_d due to the fact that T_d is significantly larger than T_{cpu}. Considering the exponential increase and the reciprocal decrease of T_{sim} with g_d, the upper bound is usually not large. Once finding it, we would search for an optimum g_d from 1 to the upper bound.

We conduct the numerical experiment to understand this with varying n and g where the period of CPU is 0.27ns and doing hash function needs 344 clock cycles. As shown in Fig. 5.2, T_{sim} increases extremely when g_d is greater than 4, and the minimum of T_{sim} can be reached by $g_d \leq 4$.

After the three steps above, we can obtain f_{sg}, f_i, and g_d that would minimize the overall execution time of M²ID. Let us take an example to interpret M²ID. As shown in Fig. 5.3, there exists $g = 3$ items each being attached by 3 tags. Using the parameter configuration method presented in this subsection, we have $f_{sg} = 9$ and $g_d = 1$. The tags are divided into 3 segments with s_{sg} (c.f. Fig. 5.3a) and then the reader finds the proper seed s_i of the i-th segment (c.f. Fig. 5.3b). Finally, each tag calculates its position with the received parameters and decides whether/when to respond following (5.2). If recognizes itself as a representative

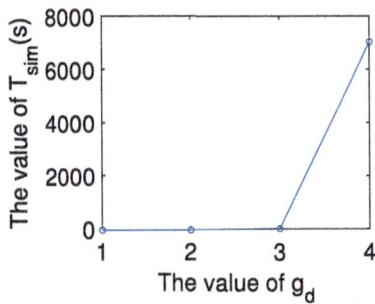
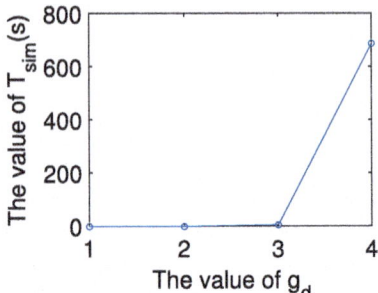

(a) The value of T_{sim} when $n = 1000$ and $g = 100$.

(b) The value of T_{sim} when $n = 1000$ and $g = 200$.

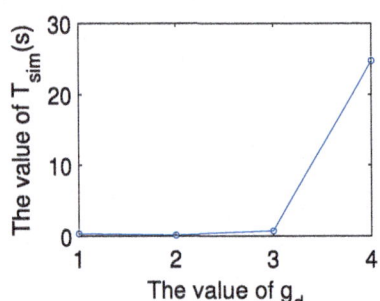
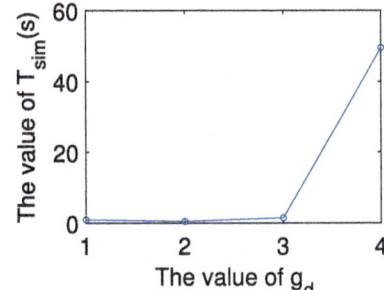

(c) The value of T_{sim} when $n = 1000$ and $g = 500$.

(d) The value of T_{sim} when $n = 2000$ and $g = 1000$.

Fig. 5.2 The impact of g_d on T

tag, the tag replies in the calculated position in the response frame. Consequently, only the representative tags respond in sequence (c.f. Fig. 5.3c).

As mentioned in Sect. 5.3.1, we assume the downlink works in the error-free channel and the uplink works in the unstable channel. Hence, we assume the probability of each received bit from the tags occurring bit inversion is p_e. The probability of a missing representative tag which is undetected in M²ID resulting from the bit inversion is written as

$$P_{unde} = 1 - P_m (1 - p_e) = p_e, \tag{5.14}$$

where P_m is the probability that a representative maps to a singleton slot and the value is 1. Therefore, the probability of detecting the real missing representative tag is expressed as

$$P_{cd} = 1 - P_{unde}^{M_a} = 1 - p_e^{M_a}, \tag{5.15}$$

where the M_a is the detection threshold. Therefore, the reliability of M²ID working in the unstable channel is estimated.

5.5 M²ID+: The Improvement of M²ID

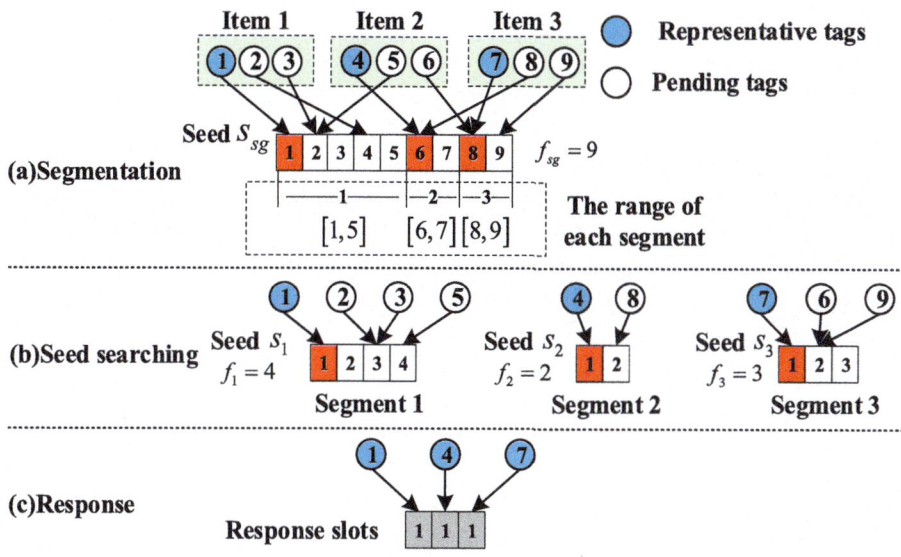

Fig. 5.3 An illustration of M²ID when $n = 9$ and $g = 3$

5.5 M²ID+: The Improvement of M²ID

In this section, we introduce M²ID+ to more actively trade-off the computation time and the communication time to further improve the time efficiency.

5.5.1 Motivation

M²ID can effectively complete the missing multi-tagged item event detection, its performance, however, is hindered by the time cost of the reader-tag communication, as described below: First, probing all representative tags is time-consuming because the detection time cost is proportionate to their size while it is adequate for the probabilistic detection to interrogate part of the representative tags with the given reliability requirement. Second, the equation (5.3) indicates that the frame size in the segmentation f_{sg} increases as the system scales up, the time spent on broadcasting boundary values of each segment will thus soar following (5.7), degrading the time efficiency. Moreover, we have to broadcast the different frame sizes used for the seed searching of each segment in M²ID, which introduces extra time costs.

To tackle the drawbacks of M²ID, we introduce the following three approaches to further embrace the computation-communication trade-offs, improving the time efficiency:

- We select the partial tags via sampling instead of the entire tag set in M²ID, which reduces the number of representative tags participating in the detection. Note that if both the sampling ratio and the required reliability are 1, M²ID+ can achieve the deterministic detection.
- We avoid spending too much time broadcasting the boundary values of each segment and improve the time efficiency via two steps. (1) We divide the sampled tags into categories and each is further segmented into multiple segments with a smaller frame size for the subsequent segmentation. (2) We introduce the seed searching into the segmentation instead of selecting an arbitrary seed in M²ID, which would further reduce the frame size used in the segmentation.
- During the seed searching of each segment, we set the identical frame size across all segments of a category so that the readers broadcast its value once instead of multiple times in M²ID.

We would like to explain that M²ID+ brings extra computation time cost compared to M²ID, but this reduces more communication time cost and thus the overall time cost. Let's take an example to make the explanation. Consider a system input of $g_d = 1, g = 50, n = 500$, M²ID would use 1212-bit binary sequence to broadcast the boundary values of all segments. In contrast, this cost reduces to 411 in M²ID+ where the tags are classified into 10 categories before the segmentation, including 300 bits representing the boundary values, 100 bits expressing the ten 10-bit seed sequences used in the subsequent segmentation for all categories, and about 11-bit cost equivalently to the computation time of $279\mu s$.

5.5.2 Protocol Description

In M²ID+, the reader works in two modes, i.e., offline or online. In the offline mode, the reader will conduct the parameters configuration, the operations of the sampling, and the classification according to the user's requirements. They are done in the reader once the system is confirmed before the detection. On the other side, online mode will search the seed and communicate with the tags for each executed detection. Therefore, the time cost for the detection discussed in this chapter refers to the cost of online mode.

(1) Sampling-classification: In the offline mode before the detection, the reader first picks up two arbitrary seeds and two corresponding thresholds to conduct the sampling and the classification. For example, an arbitrary tag's hashing value is $h_{s,j} = \left(H\left(ID_j, s_{sample}\right) \mod f_{sample}\right) + 1$ with the sampling seed s_{sample} and the frame size f_{sample}. The threshold Th is $\lceil p_{sample} f_{sample} \rceil$. If $h_{s,j} \leq Th$, this tag is sampled. Consequently, np_{sample} of the tags are sampled to participate in the sequent operations. The reader then makes these sampled tags do another hashing operation with the second seed s_{class} and its corresponding classification size f_{class}, i.e., $h_{c,j} = \left(H\left(ID_j, s_{class}\right) \mod f_{class}\right) + 1$.

5.5 M²ID+: The Improvement of M²ID

The value of $h_{c,j}$ indicates the category the tag belongs to. For example, If $h_{c,j} = u$, it would be classified into the category u for $u = 1, 2, ..., f_{class}$.

(2) Segmentation of a category: During the online mode for the detection, consider an arbitrary category u, the reader first divides it into multiple segments. Specifically, the reader searches for such a seed s_{seg_u} that all representative tags of this category map to unique hashing values with the frame size f_{seg_u}. Furthermore, we only record the length of each segment (i.e., $d_{q_u} = b_{eq_u} - b_{tq_u} + 1$, where q_u represents the q-th segment in the category u) instead of the boundary values in M²ID. The reader then finds another proper seed for the representative tags in each segment ensuring that their hashing values fall into $[1, g_d]$. The process of the seed searching in each segment is similar with M²ID. The difference here lies in that we use an identical frame size f_{s_u} across all segments instead of the different frame sizes in M²ID.

(3) Parameters broadcasting: The reader first broadcasts the seeds, the frame sizes, and the thresholds used in sampling and classification. For the category u, the reader broadcasts the parameters used in the segmentation, namely the frame size f_{seg_u} and the seed s_{seg_u}. The reader then sends the identical frame size f_{s_u} used in the seed searching for each segment and each segment's length d_{q_u} and the found seeds. Sequentially, the reader interrogates the sampled representative tags in the category u and waits for their responses. Repeating these operations for all categories, the reader checks the observed response slots of the representative tags. It can detect missing items if the predicated busy slots turn out to be empty.

On the tag side, each tag does a hash function to check whether it is sampled according to the received parameters. Only a sampled tag determines which category it belongs to, while the unsampled tags will keep silent. After knowing its category, the tag computes its segment and checks whether it is a representative tag. Each representative tag will then respond at a corresponding slot as M²ID does.

M²ID+ makes the sampled representative tags map to singleton slots and then identifies when all proper seeds are sought out. The key left is to configure the parameters used in the sampling, the classification, and the segmentation to minimize the overall execution time.

5.5.3 Parameter Setting

We here introduce how to set the parameters used in M²ID+ so that the detection reliability can be satisfied while the overall execution time can be minimized. Consider we have U categories (i.e., $U = f_{class}$), and the expected number of the tags and the representative tags in each category is n_s/U and g_s/U respectively under the sampling ratio p_{sample}, where $n_s = np_{sample}$, $g_s = gp_{sample}$. In an arbitrary category u, the reader divides the tags into several segments with the frame size f_{seg_u} and each segment contains g_d representative tags. Recall Sect. 5.5.2, the expected overall time cost of M²ID+ is

$$T_{total} = T_{r\_whole} + T_{res} = UT_u + g_s T_{tag}, \tag{5.16}$$

where T_{r_whole} is the expected time cost used by the reader to complete the computation and the broadcasting for U categories. T_u is the expected time cost of the category u, and T_{res} is the time duration of the response frame determined by the number of the sampled representative tags g_s.

For the u-th category, the expected time cost can be divided into the following three parts:

$$T_u = T_{seg_u} + T_{iden_u} + T_{ssb_u}, \tag{5.17}$$

where T_{seg_u} is the expected time cost used for the segmentation and its parameters transmission, T_{ssb_u} is the expected time cost of the seed searching and the seed transmission for all segments, and T_{iden_u} is the cost of broadcasting the identical frame size f_{s_u} for all segments expressed as $T_{iden_u} = L(f_{s_u})T_d$.

As mentioned above, the time cost of the segmentation contains the cost of the seed searching and the parameters broadcasting. Thus, T_{seg_u} is written as

$$T_{seg_u} = \frac{C_u n_s N_c}{U} T_{cpu} + \left(\frac{g_s}{g_d U}L(D) + L(f_{seg_u}) + L(C_u)\right) T_d, \tag{5.18}$$

where D is the expected length of each segment in all categories, and C_u is the expected round needed to find a seed for the segmentation. In addition, the value of the seed equals to C_u. Specifically, the expected length of each segment is

$$D = \mathbb{E}[d_{q_u}] = \frac{f_{seg_u} g_d U}{g_s}. \tag{5.19}$$

And the expected number of rounds is

$$C_u = \frac{1}{\prod_{q=0}^{\frac{g_s}{f_{seg_u}}-1} \frac{f_{seg_u}-q}{f_{seg_u}}} = \frac{f_{seg_u}^{\frac{g_s}{U}}}{\prod_{q=0}^{\frac{g_s}{U}-1} f_{seg_u}-q}, \tag{5.20}$$

where f_{seg_u} should meet $f_{seg_u} \geq \frac{g_s}{U}$ in order to make all representative tags map to the unique hashing values.

Moreover, T_{ssb_u} is the sum of the cost of the seed searching for all segments T_{ss_u} and the cost of broadcasting the proper seeds T_{ssb_u}. Specifically, T_{ss_u} can be written as

$$T_{ss_u} = \sum_{q=1}^{r_u} \frac{n_{q_u} N_c T_{cpu}}{p_{sms}(g_d, n_{q_u}, f_{s_u})} = \frac{n_s N_c T_{cpu}}{U p_{sms}\left(g_d, \frac{g_d n}{g}, \frac{g_d n}{g}\right)}, \tag{5.21}$$

where r_u is the number of the segment in the category u and its expected value is $\frac{g_s}{g_d U} \cdot n_{q_u}$ is the number of the tags in the q-th segment of the category u and its expected value is $\frac{g_d n}{g}$.

5.5 M²ID+: The Improvement of M²ID

The probability $p_{sms}\left(g_d, \frac{g_d n}{g}, \frac{g_d n}{g}\right)$ of finding a proper seed with the identical frame size $\frac{g_d n}{g}$ can be written as

$$p_{sms}\left(g_d, \frac{g_d n}{g}, \frac{g_d n}{g}\right) = \frac{g_d!}{g_d^{g_d}}\left(\frac{g}{g_d n}\right)^{g_d}\left(1 - \frac{g}{g_d n}\right)^{\frac{g_d n}{g} - g_d}. \tag{5.22}$$

Correspondingly, the expected time cost T_{ssb_u} of broadcasting the proper seeds of all segments in the category u is

$$T_{ssb_u} = \frac{g_s}{g_d U} L \left(\frac{1}{p_{sms}\left(g_d, \frac{g_d n}{g}, \frac{g_d n}{g}\right)}\right) T_d. \tag{5.23}$$

Therefore, the (5.17) can be rewritten by substituting with (5.18) (5.21) and (5.23):

$$T_u = \frac{n_s N_c}{U}\left(C_u + \frac{1}{p_{sms}\left(g_d, \frac{g_d n}{g}, \frac{g_d n}{g}\right)}\right)T_{cpu} + \left(\frac{g_s}{g_d U}L(D) + L(f_{seg_u}) + L(C_u) + L\left(\frac{g_d n}{g}\right)\right)T_d$$

$$+ \frac{g_s}{g_d U} L\left(\frac{1}{p_{sms}\left(g_d, \frac{g_d n}{g}, \frac{g_d n}{g}\right)}\right)T_d. \tag{5.24}$$

Recall the operation of $L(\cdot)$, we make the following settings for the analysis feasibility: $\max\{D\} = 3 f_{seg_u} g_d U/g_s$, $\max\{n_{q_u}\} = 3 g_d n/g$, $\max\{f_{seg_u}\} = f_{seg_u}$, $\max\{C_u\} = C_u$. Thus, the expression (5.24) can be expanded as

$$T_u^* = \frac{n_s N_c}{U}\left(C_u + \frac{1}{p_{sms}\left(g_d, \frac{g_d n}{g}, \frac{g_d n}{g}\right)}\right)T_{cpu} + \left(\log_2(f_{seg_u}) + \log_2(C_u) + \log_2\left(\frac{g_d n}{g}\right)\right)T_d$$

$$+ \frac{g_s}{g_d U}\log_2\left(\frac{3 f_{seg_u} g_d U}{g_s p_{sms}\left(g_d, \frac{3g_d n}{g}, \frac{g_d n}{g}\right)}\right)T_d. \tag{5.25}$$

And we thus have the expected overall execution time of all categories at the side of the reader as follows:

$$T_{r\_whole} = \sum_{u=1}^{U} T_u^* = U T_u^* = \left(C_u + \frac{1}{p_{sms}\left(g_d, \frac{g_d n}{g}, \frac{g_d n}{g}\right)}\right) n_s N_c T_{cpu}$$

$$+ \left(\log_2(f_{seg_u}) + \log_2(C_u) + \log_2\left(\frac{g_d n}{g}\right)\right) U T_d + \frac{g_s}{g_d}\log_2\left(\frac{3 f_{seg_u} g_d U}{g_s p_{sms}\left(g_d, \frac{3g_d n}{g}, \frac{g_d n}{g}\right)}\right) T_d. \tag{5.26}$$

Therefore, the expected overall execution time of our proposed M²ID+ is expressed as

$$T_{total} = \left(C_u + \frac{1}{P_{sms}\left(g_d, \frac{g_d n}{g}, \frac{g_d n}{g}\right)}\right) n_s N_c T_{cpu} + \left(\log_2(f_{seg_u}) + \log_2(C_u) + \log_2\left(\frac{g_d n}{g}\right)\right) U T_d$$

$$+ \frac{g_s}{g_d} \log_2\left(\frac{3 f_{seg_u} g_d U}{g_s P_{sms}\left(g_d, \frac{3 g_d n}{g}, \frac{g_d n}{g}\right)}\right) T_d + g_s T_{tag}. \tag{5.27}$$

Obviously, T_{total} is determined by p_{sample}, U, f_{seg_u}, and g_d. The key of achieving the best time efficiency is to minimize T_{total} with the optimum values of these four parameters.

To this end, we first show the configuration of the sample ratio p_{sample} under the requirement of the detection. Second, we discuss the optimum number of the categories U, the optimum frame size for the segmentation f_{seg_u} in the category u, and the optimum number of the representative tags in each segment g_d. Third, we derive the identical frame size for the specific situation after the classification and the segmentation.

(1) The optimum sampling ratio: A smaller sampling ratio usually yields less time cost, but the unbounded decreasing would make the detection unreliable. Consequently, we must set an appropriate sampling ratio. The correct detection probability in the unstable channel is expressed as

$$P_d = 1 - \left(1 - p_{sample} P_m (1 - p_e)\right)^{M_a}, \tag{5.28}$$

where P_m is the probability of an arbitrary representative tag mapping to a singleton slot in the response frame, p_e is the probability of the bit inversion induced by the unstable channel, and M_a is the threshold. Recall Sect. 5.5.2, we ensure the probability of an arbitrary representative tag mapping to a singleton slot in the response frame is 1. Consequently, given the reliability requirement α and establishing $P_d \geq \alpha$, we have

$$p_{sample} \geq \frac{1}{1 - p_e}\left(1 - (1 - \alpha)^{\frac{1}{M_a}}\right), \tag{5.29}$$

which is the sampling ratio we need.

(2) Configuring the number of the categories U, the frame size used for the segmentation f_{seg_u}, and the number of the representative tags in each segment g_d: Recall (5.27), the object moves to configure the number of the categories U, the number of the representative tags g_d in each segment, and the frame size of the segmentation f_{seg_u} in each category to minimize the execution time T_{total}. Because it is difficult to directly derive them, we have to search for the optimum value of these parameters.

Let us start with the g_d. We have shown in Sect. 5.4.4 that g_d would be small with a high probability because of the exponentially increasing rate. We set the range of g_d as $1 \leq g_d \leq 4$. And U should be set to ensure that each category contains the representative tag(s), we thus have $1 \leq U \leq \frac{g_s}{g_d}$. The lower-bound of f_{seg_u} is set as $\frac{g_s}{U}$ according to the

5.5 M²ID+: The Improvement of M²ID

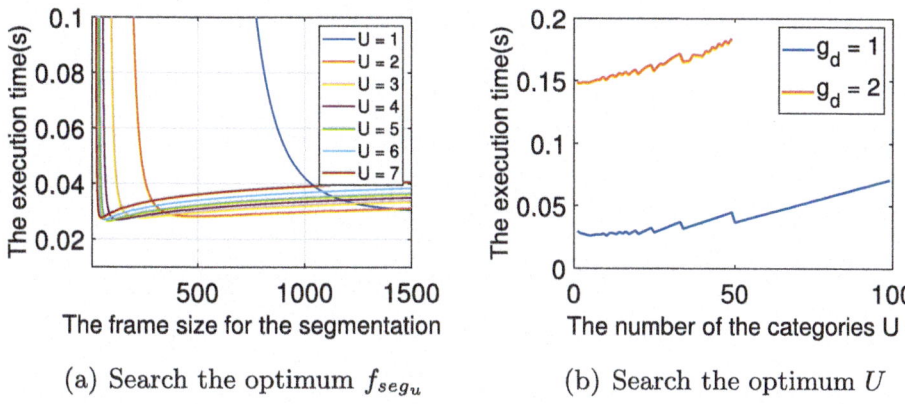

(a) Search the optimum f_{seg_u} (b) Search the optimum U

Fig. 5.4 The parameters configuration when $n = 1000$ and $g = 100$

(5.20). The upper bound f_{seg_u} can be set if f_{seg_u} makes $C_u(f_{seg_u}, U) < 2$, which means that the expected round of searching a proper seed for the segmentation is less 2.

Now, that the range of these three parameters has been determined, we start to seek out the proper values. We will take an example to explain the searching process with a system consisting of 100 items and 10 tags on each item (i.e., a total of 1000 tags in the system), as shown in Fig. 5.4. The period of the CPU is 0.27ns with the clock round of 344 doing the hash function and the downlink transmission rate is 40.97kb/s. The curves in Fig. 5.4a show the time cost with the different number of categories and the different frame size for the segmentation. Note that it is adequate to show $1 \leq U \leq 7$ because the execution time soars for $U > 6$. More specifically, the curves at the different g_d in Fig. 5.4b show the time cost with the different number of categories under the corresponding optimum f_{seg_u}. Hence, we can obtain the optimum parameters as follows: The number of the representative tags in each segment $g_d^* = 1$, the frame size $f_{seg_u}^* = 35$ used for the segmentation of a category u, and the number of the categories $U^* = 5$.

As mentioned in Sect. 5.5.2, the optimum parameter configuration for the sample, the classification, and the segmentation can be calculated offline and can be recorded in the reader's storage. The reader will select the proper configuration from the storage once the system input is fixed.

(3) The identical frame size of the seed searching for each segment: Before the classification and the segmentation, the expected number of the tags segmented into the q-th segment of the category u is $\mathbb{E}[n_{q_u}] = ng_d/g$ under the fixed g_d. After that, the true number of the tags n_{q_u} might be different, which results in directly using the $n_{q_u} = g_d n/g$ is inefficient.

For the category u after the classification and the segmentation, the number of segments is r_u under the fixed g_d. The expected cost of the seed searching for all segments is

$$T^*_{ss_u} = \sum_{q=1}^{r_u} \frac{n_{q_u} N_c T_{cpu}}{p_{sms}(g_d, n_{q_u}, f_{s_u})}. \quad (5.30)$$

The difference f_{s_u} between (5.17) and (5.30) is that the former is an expected value used to set U, g_d and f_{seg_u} while the latter optimum based on the true value of the n_{q_u}. Now, our objective is to minimize the (5.30) with a proper $f^*_{s_u}$. We have

$$\frac{\partial T^*_{ss_u}}{\partial f_{s_u}} = \frac{N_c T_{cpu} g_d^{g_d}}{g_d!} \sum_{q_u=1}^{r_u} \frac{g_d^2 \left(1 - \frac{g_d}{f_{s_u}}\right)^{g_d - n_{q_u}}}{f_{s_u}^2 \left(\frac{g_d}{f_{s_u}}\right)^{g_d+1}} + \frac{g_d(g_d - n_{q_u})\left(1 - \frac{g_d}{f_{s_u}}\right)^{g_d - n_{q_u} - 1}}{f_{s_u}^2 \left(\frac{g_d}{f_{s_u}}\right)^{g_d}}.$$

(5.31)

The result is 0 when $f^*_{s_u} = \frac{1}{r_u}\sum_{q_u=1}^{r_u} n_{q_u}$. Thus, the proper frame size $f^*_{s_u}$ of the category u is fixed.

We next take an example to explain M²ID+. Our system is shown in Fig. 5.5. The configuration is $g_d = 1$, $U = 2$ and all tags are sampled. Then, the tags in the category 1 are divided into 5 segments with the $f_{seg_1} = 7$ and the reader searches the proper seed for each segment with the identical frame size $f_{s_1} = 3$. Finally, only the representative tags respond

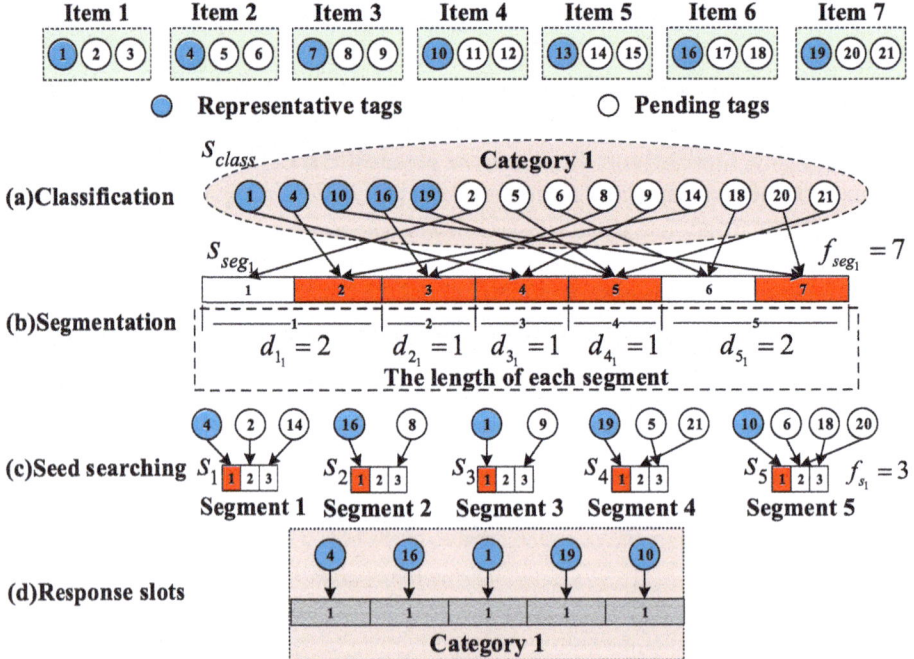

Fig. 5.5 Illustrating M²ID+ with 7 items each attached by 3 tags

for the missing item detection, and all the slots in the response frame are singleton, that said the utilization ratio of the slots to identify the representative tags reaches 100%. The tags in the category 2 would repeat the above process.

5.6 Performance Evaluation

We evaluate the performance of the proposed M^2ID and M^2ID+ in terms of the detection probability and the execution time, and compare their performance with the most related work named BPI [11] that uses bloom filter to identify the tags in the multi-tagged RFID systems. The timing parameters in the simulation follow the EPC-global Gen2 standard [13]. Specifically, the transmission rate is 40.97kb/s, and a broadcast slot and a response slot are $T_d = 24.4\mu s$ and $T_{tag} = 290.8\mu s$, respectively. Furthermore, the number of the rounds accomplishing once hash function is $N_c = 344$ [15] and the period of the CPU's clock is $T_{CPU} = 0.27$ns. The parameters like the sampling ratio and the number of representative tags in each segment are set from the analysis. The results are obtained from 1000 independent runs.

Performance Verification: We evaluate the proposed protocols under five scenarios. In the simulation, the threshold of the missing items is set to $M_a = 2$ and the required detection reliability varies from $\alpha = 95\%$ to $\alpha = 99\%$ in the first two scenarios and is fixed to $\alpha = 95\%$ in the third scenario. In the fourth scenario, we set $\alpha = 100\%$ and $p_{sample} = 100\%$ to enable the tag identification of M^2ID+, and we show the superior time efficiency of the proposed protocols compared with the state-of-the-art BPI [11]. The above simulations work in the ideal channel, i.e., $p_e = 0$. Therefore, the detection probability equals to the ratio of the correct detection. In the last simulation, we verify the effectiveness of the proposed protocols under the unstable channel, and the ratio of the correct detection is set as $\alpha = 95\%$.

(1) In the first scenario, there exists 10 tags on each item and the number of the tags varies from 1, 000 to 5, 000. The simulation results of the detection probability and the execution time are depicted in Fig. 5.6. The results show that the proposed M^2ID and M^2ID+ can satisfy the requirement of the detection reliability. Yet, these two protocols have to spend more time on finding a missing item event as the number of the tags in the system increases when there would be more representative tags leading to the longer time cost of the seed searching and the parameters broadcasting. We can also observe that M^2ID+ is more time-efficient. As shown in Fig. 5.6c, M^2ID+ is faster 3x than M^2ID when the number of the total tags is 5000.

(2) In the second scenario, we investigate the impact of the number of tags on one item on the detection probability and the execution time. To this end, we set the total number of the tags in the system as 1, 000, and vary the number of the tags on each item from 2 to 10. We can draw from Fig. 5.7 similar conclusions as in the first scenario that both M^2ID and M^2ID+ can achieve the required reliability, but the latter is more time-efficient and the gain is 2x at least.

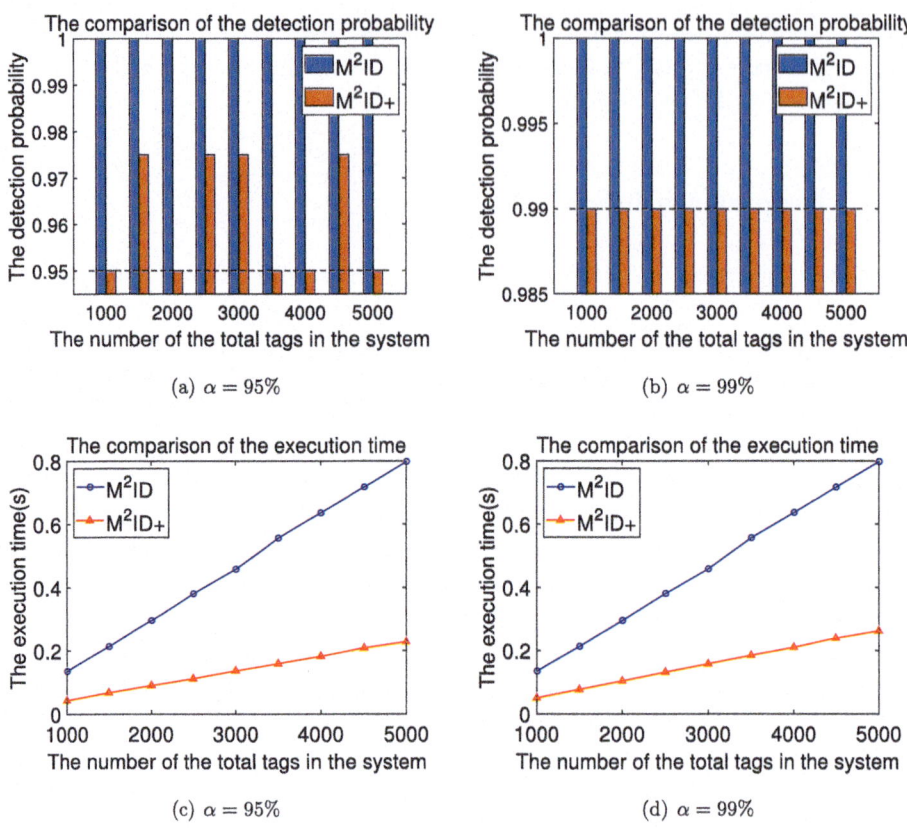

Fig. 5.6 The detection probability and the execution time versus the number of the total tags with $\alpha = 95\%$ and $\alpha = 99\%$

(3) In the third scenario, we show the impact of the system scale on time efficiency. The experiment consists of two cases: The first case witnesses 10 tags on each item while the number of the total tags varies from 5,000 to 30,000; The second case has 30,000 tags while the number of the tags on each item changes from 2 to 10. They indicate the change of the number of items. We can observe from Fig. 5.8 that both M^2ID and M^2ID+ spend more time as the system scales up, but the increasing speed of M^2ID+ is significantly slower.

(4) In the fourth scenario, we compare the proposed protocols with the state-of-the-art solution BIP. To this end, we set the sample ratio in M^2ID+ to 100%. Figure 5.9a shows that M^2ID+ is most time-efficient and M^2ID+ is faster than BIP when the number of the tags with the change of the item population. Similarly, Fig. 5.9b depicts the execution time when the total number of the tags is 30000 and the number of the tags on each item varies from 2 to 10. Similarly, M^2ID+ spends least time among these three protocols. From Fig. 5.9,

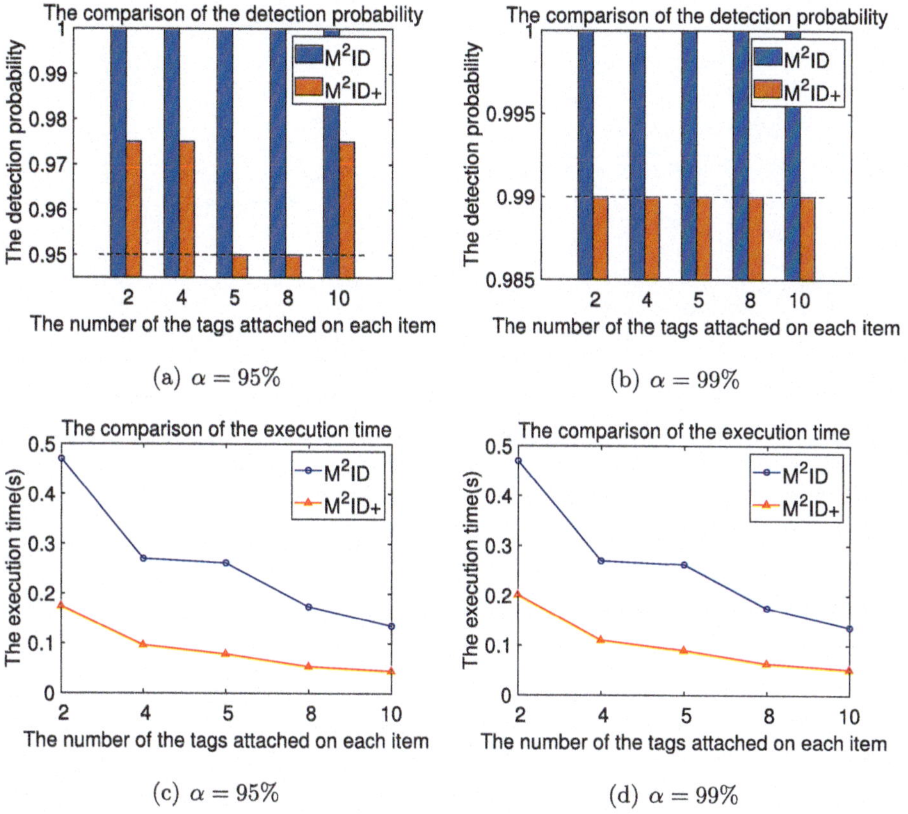

5.6 Performance Evaluation

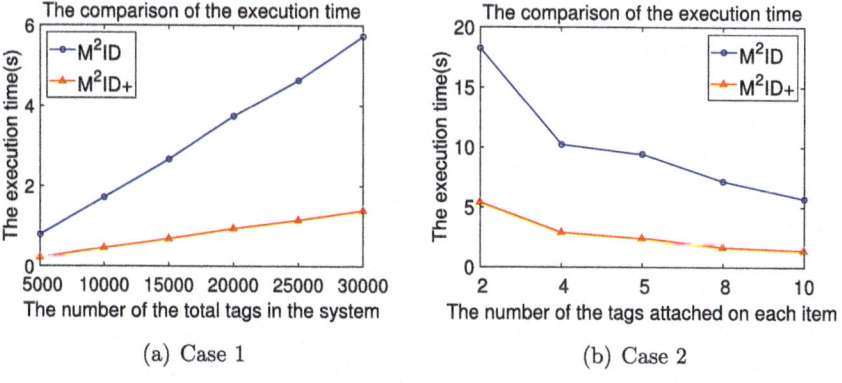

Fig. 5.7 The detection probability and the execution time versus The number of the tags on each item with $\alpha = 95\%$ and $\alpha = 99\%$

Fig. 5.8 The time efficiency versus the system scale under $\alpha = 95\%$: **a** Case 1 with the number of the total tags varied from 5000 to 30000 and 10 tags on each item; **b** Case 2 with the number of the tags on each item varied from 2 to 10 and 30000 tags in total

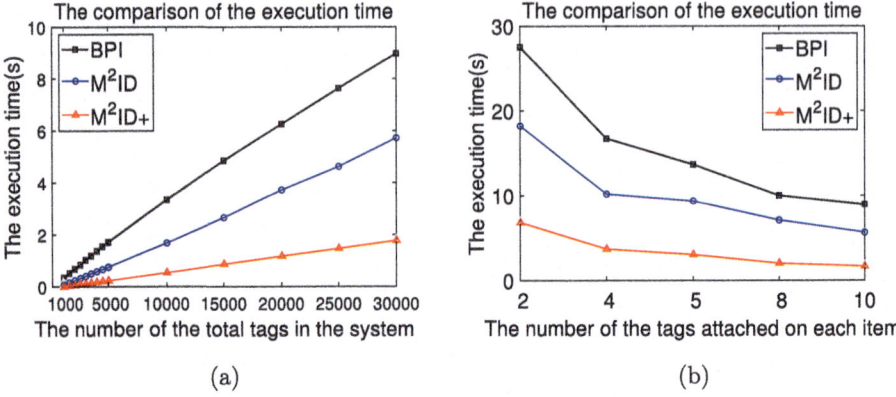

Fig. 5.9 a The execution time with the number of the total tags varied from 1000 to 30000 and the number of the tags on each item is set to 10. **b** The execution time with the number of the tags on each item varied from 2 to 10 and the number of the total tags is set to 30000

M^2ID+ is still the most effective among them and M^2ID+ achieves at least 4x performance gain compared with BPI.

(5) In the fifth scenario, we verify the effectiveness of our proposed protocols in the unstable channel where the downlink works in error-free and the probability of bit inversion in the uplink is from $p_e = 0.1$ to $p_e = 0.2$. Figure 5.10 illustrates that 10 tags on each item while the number of the total tags varies from 1,000 to 5,000 with different p_e. The ratio of the correct detection degrades in M^2ID but still satisfies the requirement of the correct detection. The execution time of M^2ID+ grows with the increasing p_e since M^2ID+ has to increase the sample ratio to guarantee the correct detection.

5.7 Conclusion

This chapter has addressed a variation on the missing item event detection problem arising from multi-tagged items in RFID systems. The application of the prior works to the new problem suffers low time efficiency due to repeated checks of one item. To overcome this drawback, we have provided two solutions, namely M^2ID and M^2ID+, from the perspective of the trade-off between the computation and the communications. They have used the seed selection to ask a subset of the tags in systems to report their presence while M^2ID+ can achieve both probabilistic detection and deterministic identification. We have also derived the optimum parameters under the unstable channels. The simulation results have confirmed the superiority in terms of time efficiency under the required detection reliability to the existing state-of-the-art solution.

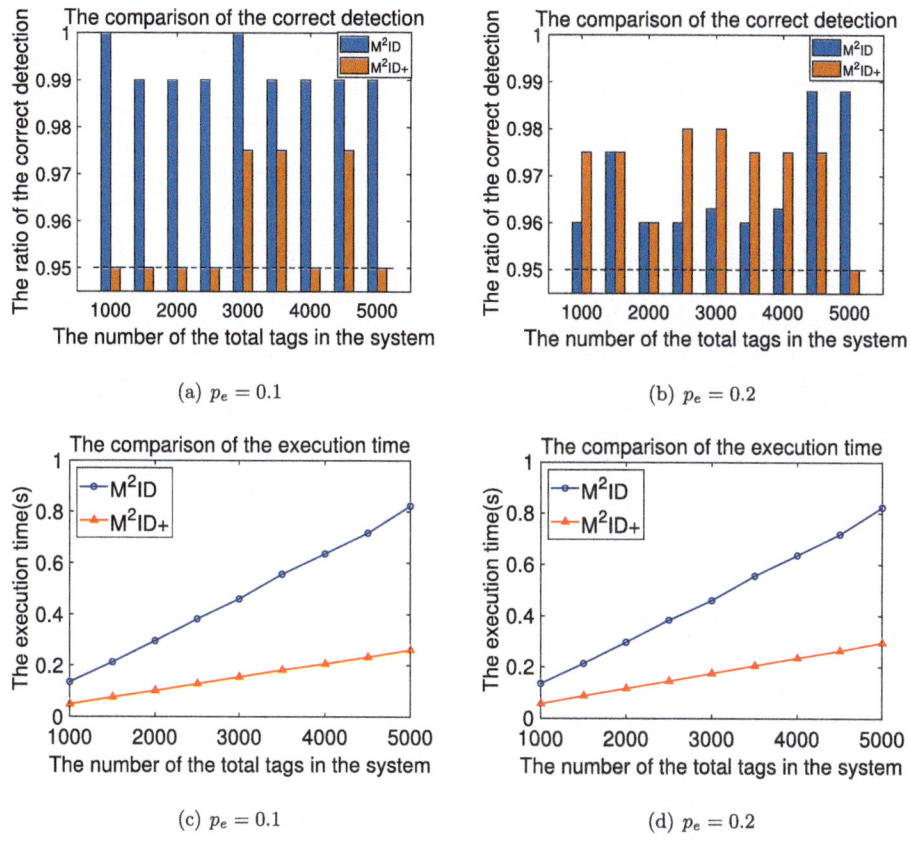

Fig. 5.10 The ratio of the correct detection and the execution time versus the error probability of the uplink with $p_e = 0.1$ and $p_e = 0.2$

References

1. C. C. Tan, B. Sheng, Q. Li, How to monitor for missing rfid tags, in *IEEE ICDCS* (2008), pp. 295–302
2. W. Luo, S. Chen, T. Li, Y. Qiao, Probabilistic missing-tag detection and energy-time tradeoff in large-scale rfid systems, in *ACM MobiHoc* (2012), pp. 95–104
3. W. Luo, S. Chen, Y. Qiao, T. Li, Missing-tag detection and energy-time tradeoff in large-scale rfid systems with unreliable channels. IEEE/ACM ToN **22**(4), 1079–1091 (2014)
4. M. Shahzad, A. X. Liu, Expecting the unexpected: Fast and reliable detection of missing rfid tags in the wild, in *IEEE INFOCOM* (2015), pp. 1939–1947
5. J. Yu, L. Chen, R. Zhang, K. Wang, Finding needles in a haystack: Missing tag detection in large rfid systems. IEEE TCOM **65**(5), 2036–2047 (2017)

6. J. Yu, L. Chen, R. Zhang, K. Wang, On missing tag detection in multiple-group multiple-region rfid systems. IEEE TMC **16**(5), 1371–1381 (2016)
7. J. Yu, W. Gong, J. Liu, L. Chen, K. Wang, R. Zhang, Missing tag identification in cots rfid systems: Bridging the gap between theory and practice. IEEE Trans. Mobile Comput. **19**(1), 130–141 (2019)
8. T. Li, S. Chen, Y. Ling, Identifying the missing tags in a large rfid system, in *Proceedings of the eleventh ACM international symposium on Mobile ad hoc networking and computing* (ACM, 2010), pp. 1–10
9. R. Zhang, Y. Liu, Y. Zhang, J. Sun, Fast identification of the missing tags in a large rfid system, in *2011 8th Annual IEEE Communications Society Conference on Sensor, Mesh and Ad Hoc Communications and Networks* (IEEE, 2011), pp. 278–286
10. X. Liu, K. Li, G. Min, Y. Shen, A.X. Liu, W. Qu, Completely pinpointing the missing rfid tags in a time-efficient way. IEEE ToC **64**(1), 87–96 (2015)
11. X. Xie, X. Liu, H. Qi, K. Li, Fast identification of multi-tagged objects for large-scale rfid systems. IEEE Wireless Commun. Lett. **8**(4), 992–995 (2019)
12. J. Yu, W. Gong, J. Liu, L. Chen, K. Wang, On efficient tree-based tag search in large-scale rfid systems. IEEE/ACM ToN **27**(1), 42–55 (2019)
13. *EPC radio-frequency identity protocols Class-1 Generation-2 UHF RFID Protocol for communication at 860 MHz - 960 MHz*. EPC, 2.0.1 ed., (2015)
14. J. Yu, P. Zhang, L. Chen, J. Liu, R. Zhang, K. Wang, J. An, Stabilizing frame slotted aloha based iot systems: A geometric ergodicity perspective. IEEE JSAC **39**(3), 1–12 (2020)
15. M. O'Neill et al., Low-cost sha-1 hash function architecture for rfid tags. RFIDSec **8**, 41–51 (2008)

Practical Hashing-Free Access Implementation with COTS RFID Systems

This chapter presents schemes for detecting all missing tags in COTS RFID systems. Prior works on missing tag detection have predominantly relied on hash functions implemented at individual tags. However, in practice, COTS RFID tags do not support hash functions. To bridge this gap between theoretical approaches and practical implementation, this chapter focuses on detecting missing tags with COTS Gen2 devices. We first introduce a point-to-multipoint protocol, named P2M, which operates within an analog frame-slotted Aloha paradigm to interrogate tags and collect their electronic product codes (EPCs). A tag is identified as a missing tag if its EPC is absent from the collected data. To mitigate the time cost associated with tag response collision in P2M, we further present a collision-free point-to-point protocol, named P2P, which selectively specifies a tag to respond with its EPC in each slot. If the EPC is not received, the tag is considered missing. We have developed two bitmask selection methods to enable selective querying while minimizing communication overhead. Both P2M and P2P protocols have been implemented using COTS RFID devices, and their performance has been evaluated under various settings.

Chapter roadmap: The rest of this chapter is organized as follows. Section 6.1 highlights the importance of studying missing tag identification with COTS Gen2 devices. Section 6.3 reviews the system model and the problem statement. In Sect. 6.4, we introduce the P2M on Gen2-compatible protocol to identify missing tags, estimate the time consumption, and induct the parameter configuration. The P2P to singularize tags in every slot, avoiding collision events while improving time efficiency, is proposed in Sect. 6.5. In Sect. 6.6, we evaluate the performance of the proposed missing tag identification protocols with the COTS Gen2 devices. Section 6.2 briefly summarizes the existing missing tag monitoring protocols. Finally, we summarize the contributions of this chapter in Sect. 6.7.

6.1 Introduction

To enable worldwide commercial implementation of RFID, the EPC-global, an organization that was formed in 2003, developed the Gen2 air interface protocol [1] for ultra-high-frequency (UHF) RFID systems. This protocol has been adopted as the ISO 18000-6C standard and has become mainstream specification worldwide for commercial off-the-shelf (COTS) RFID devices like ImpinJ [2] and ThingMagic series [3]. A Gen2 RFID system comprises two types of devices: passive RFID tags and RFID reader. A passive tag is a lightweight battery-free device that can record information of a physical object and is able to capture the energy in the wireless signal of its nearby RFID reader and modulate this signal by adjusting the impedance match on its antenna so that a message of zeros and ones is back-scattered to the reader.

Identifying missing tags, which is to completely pinpoint the tags that should be in the coverage range of the reader but are absent, is one of the most important RFID-enabled services. According to the statistics presented in [4], inventory shrinkage, a combination of shoplifting, internal theft, administrative and paperwork error, and vendor fraud, resulted in about 49 billion dollars in loss for retailers in 2016. In this context, RFID provides a promising technology to reduce financial loss by deploying a reader to monitor passive tags attached to products in its coverage range and conducting missing tag identification regularly to find missing items in time.

The study of missing tag identification was initiated in the research community about 10 years ago, and ever since then ten-year effort has been dedicated to reducing communication overhead, producing a large body of work. However, none of the previous work is compatible with the Gen2 standard so they cannot be implemented in practice, which leaves billions of deployed COTS tags behind. The failure of the prior work mainly results from the two reasons:

1. *Hashing-dependent slot selection*: Prior work on missing tag identification requires the functionality of hashing in tags so that each tag can select and respond in a random but predictable slot corresponding to the hash value of its electronic product code (EPC) and a random seed. While the hashing functionality has never been implemented in any COTS tags high energy consumption and manufacturing costs will be incurred otherwise (e.g., over 1,000 gate equivalents for hardware), which is contradictory to what is expected of RFID.
2. *Complete visibility for slot states*: Prior work must definitely know the states of each slot, e.g., empty and busy, which depends on the number of one-bit responses from tags in this slot and exploits the empty slots that should be busy to identify missing tags. While a COTS Gen2 reader only reports successful reads in a time interval, disabling the utility of empty slots. Hence, the previous work cannot be implemented in COTS RFID devices.

Motivated by the observations above, we argue that a systematical study on missing tag identification with COTS Gen2 devices is called for to maximize the function of widely deployed Gen2 RFID systems and to reduce financial losses.

In this chapter, we develop two protocols named point-to-multipoint (P2M) and point-to-point (P2P). We implement P2M and P2P in extensive scenarios using COTS Gen2 devices: one ImpinJ reader and 20 ImpinJ Monza tags. The results show that P2P achieves time efficiency gains of about 4x and 6x over P2M on average in the identification of all missing tags and the detection of the first missing tag. We also confirm the correctness of bitmask selection approaches of P2P in larger systems.

6.2 Related Work

In this section, we briefly summarize the existing missing tag monitoring protocols that can be classified into two categories: probabilistic detection and deterministic identification.

Probabilistic missing tag detection: This type of protocol detects a missing tag event with a predefined probability. Tan et al. initiate the study of missing tag detection and propose a solution called TRP in [5]. To detect a missing tag event, TRP first builds a virtual bitmap by using a hash function to predict the response slots of tags and compares it with actual slot states measured from the response of the tags in the population. If an expected busy (singleton or collision) slot turns out to be empty, then the tag(s) corresponding to this slot are regarded to be absent. Because the probability of a collision slot having only missing tags is very low when the missing tag size is small, collision slots are less useful than singleton ones. Given the importance of singleton slots, follow-up works [6, 7] employ multiple seeds to tune empty and collision slots to singleton slots, which increases the detection probability and thus improves time efficiency. Subsequently, the existence of unknown tags would make an empty slot a missing tag mapped to become busy and will interfere with the detection. To deal with the interference, the work [8] and Yu et al. [9] expand the frame size in the detection with unknown tag size and design Bloom filter from the known tags to depress the unknown ones, respectively. Consider a different kind of application scenario, Yu et al. [10] designed several Bloom-filter-based approaches to detect missing tags in RFID systems where multi-category tags are distributed in multiple regions. More recently, Yang et al. [11] developed an on-tag hashing function that needs to write offline calculated hash values to all tags, and illustrate how to use this function to probabilistically detect missing tags.

Deterministic missing tag identification: Deterministic protocols are to exactly identify which tags are absent. Li et al. develop a series of identification protocols in [12] to reduce the time cost step by step by reconciling 2-collision slots and iteratively deactivating the tags of which the presence has been verified, respectively. Zhang et al. propose identification protocols in [13] which store and compare the bitmaps of tag responses in all rounds and look for changes at the corresponding bits among all bitmaps to determine the present and absent tags. Liu et al. [14] essentially combine the multi-seed method in [6] with the deactive-based method in [12] to improve the identification performance. Subsequently, Liu et al. [15] further enhance the prior work by reconciling both 2-collision and 3-collision slots and filtering the empty and unreconcilable collision slots to improve time efficiency. Recently, physical-layer information is exploited to accelerate missing tag identification. Zheng et al. [16] measure changes in signal strength in each slot and models missing tag identification using Compressing Sensing, which reduces time cost toward the same order of magnitude as the missing tag population. In contrast, Chen et al. [17] uses changes in signal strength in each slot to construct a Bloom filter, which can achieve a similar time efficiency while handling an arbitrary number of missing tags.

Compared with the previous work, the novelty of this chapter lies in that we design bitmask selection methods and conduct deterministic missing tag identification using COTS RFID devices without the requirement for hash functions at tags and for writing hash values to tags.

6.3 System Model and Problem Statement

A typical Gen2 RFID system consists of a reader and multiple passive tags. The reader can charge, synchronize, and collect information from tags, while tags each having an EPC are usually attached to physical objects, producing a one-on-one map between a tag and an object. To interact with battery-free tags, the reader initially transmits a continuous wave to the tags. The tags capture energy from the incoming wave to power themselves on one hand and use this wave as a carrier to backscatter their information bits with ON-OFF keying on the other hand. Specifically, the tags send a '1' bit by adjusting the impedance match on their antennas to reflect the reader's wave and a '0' bit by remaining silent [18].

The Missing Tag Identification Problem: Consider a Gen2 system containing a reader and n tags $\{x_1, x_2, \cdots, x_n\}$ and that the reader knows all tag EPCs, there exists an event that m out of the n tags are missing due to the damage of these tags or the disappearance of their corresponding objects. *The missing tag identification problem is to exactly find the m missing tags.* In this problem, execution time which is measured as the time spent achieving the task is the most important metric. The earlier missing products are found, the more significantly financial loss is reduced.

Limitations of prior work: A large body of work is proposed to accelerate the identification process on top of the assumption that response slots of tags are predictable via hashing operations. Though the works are promising in improving time efficiency, the reality is that the widely deployed Gen2 tags cannot support the hash function that is the prerequisite of these works. Moreover, no manufacturer declares that the hash function will be packaged into commercial tags in the near future.

Why is the hashing functionality not supported by COTS tags? The main reason lies in the high energy consumption and manufacturing cost introduced by the hardware design of the hash function.[1] In particular, thousands of gate equivalents are required for current common hash functions, such as SHA-1 and SHA-256 [19] require 8,120 and 10,868 gate equivalents with power consumption $10.68\mu A$ and $15.87\mu A$, respectively. Even the most compact hash function that is presented in theory and is not available for COTS tags, e.g., PRESENT-80 [20], still requires 1,075 gate equivalents. Considering the huge market of RFID (e.g., 1.82×10^{10} tags in 2017), enabling hash function in tags will incur extremely high costs.

The proposed solutions without the requirement of the hash function. It is still an open question of how to identify all missing tags without the hash function in the Gen2 system. To bridge this gap, we design two Gen2-compatible missing tag identification protocols by using commands specified in the Gen2 standard, such as Q-command and *Select* command. As our protocols can be implemented in COTS RFID devices, they can be used to identify missing items in RFID-deployed scenarios like Walmart [21] and River Island [22], to reduce or even avoid financial loss resulting from product missing event.

In what follows, we describe P2M that behaves in a point-to-multipoint manner with the Q-command used to query all tags in the system. We then show the second work, namely P2P, which ensures point-to-point communication in each slot with an exclusive bitmask and avoids empty and collision slots.

6.4 P2M: Point-to-Multipoint Missing Tag Identification

In this section, we introduce the first Gen2-compatible protocol, the point-to-multipoint Q-query, and its application to identify missing tags, and then show the parameter configuration and time cost computation.

6.4.1 Point-to-Multipoint Q-Query

The Gen2 standard specifies how the reader interrogates tags. First, the reader sends a *Query* command to initiate the interrogation. This command contains backscatter link frequency (BLF), tag-to-reader encoding method, and a Q parameter used to specify the number of slots in this query round. With the parameter Q, each tag is able to determine its response slot by selecting a random value in $[0, 2^Q - 1)$ as its slot counter. If this counter is equal to 0, the tag replies immediately with a 16-bit random number (RN16); otherwise, it shall keep silent. Upon receiving an *RN16* from a tag, the reader transmits an *ACK* containing the decoded *RN16* to acknowledge this tag. If the tag confirms the correctness of the reader-to-tag *RN16*, the tag will backscatter its *EPC* to the reader. Subsequently, the reader issues a *QueryRep* to instruct tags to decrement their slot counters and the tags whose counters are equal to 0 reply

---

[1] Gate equivalent is a key performance metric in evaluating the efficiency and availability of a hardware design. The more gate equivalents are required, the higher the implementation overhead and cost are.

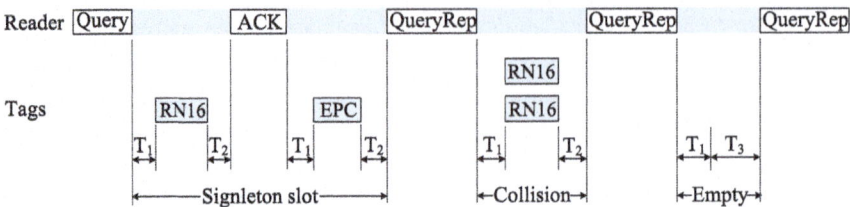

Fig. 6.1 Link timing of P2M communications. The Gen2 standard has strict requirements for each command format and link timing parameters T_1, T_2, and T_3 that stand for an interval-command time, enabling the computation of overall interrogation time

with another *RN16*, indicating the start of a new slot.[2] Figure 6.1 illustrates the *Q*-query process and shows that there is waiting time between two continuous commands like T_1, T_2, and T_3.

Since the reader can collect all EPCs of the tags present in its coverage via the *Q*-query, it compares the collected EPCs with the ones recorded in the database. If some recorded EPCs are not present in the collected EPC set, these tags are missing and can thus be identified by the reader. This comparison is conducted at the end of the *Q*-query. P2M is superior to the existing works since they need the knowledge of all slot states which cannot be obtained from a COTS reader. The main question in P2M is when the *Q*-query should be terminated.

6.4.2 Encoding Methods

The quest for low cost, tiny size, and battery-free tags severely limits their computation and hardware capabilities. It is thus important and necessary to encode and decode data in an extremely simple and robust way. In practice, the reader-to-tag symbols are amplitude-modulated pulse interval encoding (PIE) symbols which an analogy comparator is adequate to decode. As shown in Fig. 6.2, symbol '0' in PIE comprises two intervals of the same length, namely power-on and power-off (PW: pulse width). Tari (Type A reference interval) is the duration of data-0, while the duration of data-1 is as long as $x \in [0.5, 1]$ times of data-0. The Tari values can be set as 6.25, 12.5, or 25 μs corresponding to the rates 160, 80, and 40 kbps. Different from the lightweight tags, the reader has the strong decoding capacity. The Gen2 standard specifies four encoding method for the tag-to-reader link, FM0, M2 (Miller2), M4 (Miller4), and M8 (Miller8). The data rate depends on the BLF and the encoding method. For example, if BLF is 320 kHz, the data rates of FM0, M2, M4, and M8 are 320/1, 320/2=160, 320/4=80, 320/8=40 kbps, respectively.[3]

---

[2] The counter of a tag in the *Q*-query measures the number of slots before it replies, thus setting a value to a tag's counter is equivalent to assigning a slot to this tag.

[3] The reader sets and packages the parameters, including encoding type and BLF, into a query command, and sends the command to tags.

6.4 P2M: Point-to-Multipoint Missing Tag Identification

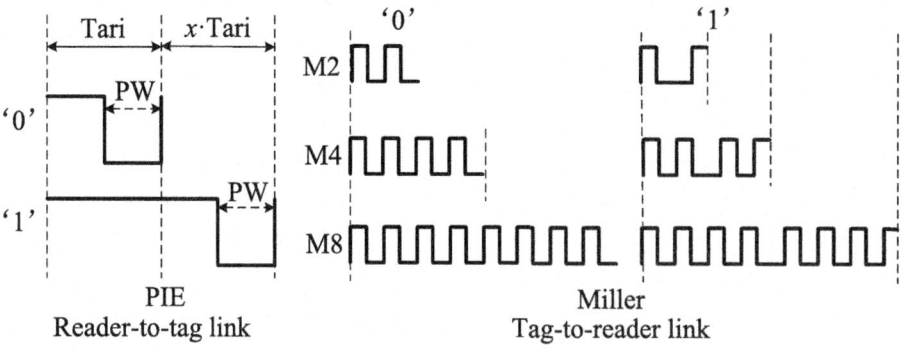

Fig. 6.2 Data encoding in the Gen2 standard

6.4.3 Configuration of the Parameter Q

From the description of the Q-query, we can observe that it is a random access process in nature, with tags randomly setting their individual counters at the beginning of the interrogation. The reader cannot predict the values picked by the tags. Consider an arbitrary slot i, there would be three states:

- If there is only one tag replying, i.e., this tag uniquely picks the value i, it is a singleton slot;
- if there are multiple tags replying, i.e., these tags pick the value i, it is a collision slot;
- if there is no tag replying, i.e., no tag selects the value i, it is an empty slot.

We make an illustration in Fig. 6.1 where one tag replies in the first slot and then two tags and no tag respond in the second and the third slots, respectively.

Among these states, only singleton slots are useful for EPC collection while collision and empty slots are useless, thus a natural optimization criterion is to ensure with a high probability that there exist n singleton slots in the interrogation, meaning that no collision occurs. Technically, the Q-query process can be formulated as the classic Ball-into-Bins problem [23]. Specifically, n tags are balls and 2^Q values (or slots) are bins. To avoid collisions with high probability, 2^Q needs to be set to $\Theta(n^2)$ [24]. Guided by this theoretical result, we set Q to $2\log n$ where log denotes the logarithm to the base 2. Under such configuration, the Q-query lasts n^2 slots.

By this setting, it is adequate for our point-to-multipoint protocol to know singleton slots, which fits well in today's COTS devices. In contrast, we note that existing works require the reader to report empty slots, which is unsupportable in the current COTS devices.

6.4.4 Calculation of the Interrogation Duration

As shown in Fig. 6.1, the three types of slots differ in their slot duration. Thus the first step in the interrogation duration computation is to figure out the number of slots in each type. Recall that we set $Q = 2 \log n$ to ensure no collision and that there are m missing tags, there would be $n - m$ singleton slots and $n^2 - n + m$ empty slots in the interrogation. As a result, the key is to compute the sizes of singleton and empty slots. To do so, we further zoom in on each slot in Fig. 6.1, and obtain the following observations:

- Singleton slot size: A singleton slot is composed of an *RN16*, an *ACK*, an *EPC*, and the inter-command time T_1 and T_2. Thus we can calculate a singleton slot size as $ACK \cdot \text{Tari} + \frac{RN16+EPC}{BLF/j} + 2(T_1 + T_2)$ where $j \in \{1, 2, 3, 4\}$ indicates different tag-to-reader encoding methods.[4]
- Empty slot size: An empty slot comprises two intervals of commands T_1 and T_3, thus its length is equal to $T_1 + T_3$.
- Inter-slot time: There is a *Query* command in the beginning of the interrogation and a *QueryRep* between any two continuous slots, so the overall inter-slot time in the whole interrogation should be $(Query + (n^2 - 1) \cdot QueryRep) \cdot \text{Tari}$.

Following these observations, now we are able to formulate the overall interrogation time of P2M is $(n - m) \cdot (ACK \cdot \text{Tari} + \frac{RN16+EPC}{BLF/j} + 2(T_1 + T_2)) + (Query + (n^2 - 1) \cdot QueryRep) \cdot \text{Tari} + (n^2 - n + m)(T_1 + T_3)$.

6.5 P2P: Point-to-Point Missing Tag Identification

Our first proposition presented previously follows the point-to-multipoint paradigm. Due to its random nature, multiple tags may reply with *RN16* simultaneously, leading to decoding failure for the reader. To deal with tag collisions, P2M sets Q to $2 \log n$, which results in considerable empty slots and wastes time. To avoid collision events while improving time efficiency, we propose P2P that performs as a point-to-point paradigm, which is able to singularize tags in every slot. As shown in Fig. 6.3, the reader cannot control the response slots of tags in P2M such that it suffers from collisions. In contrast, P2P can assign the reply order and avoid collisions, such as tags 1-5 responding in slots 1 to 5 in sequence. In what follows, we first elaborate on the missing tag identification process and then demonstrate how to build effective and efficient bitmasks.

---

[4] Either a preamble or a frame-sync will be prepended to every command, such as *RN16*, *EPC*, *ACK*, *Query*, *QueryRep* and *Select*. In addition, tags reply PC (protocol control) and CRC along with their *EPC*s. We use these commands to represent their individual length plus the extra length (bits).

6.5 P2P: Point-to-Point Missing Tag Identification

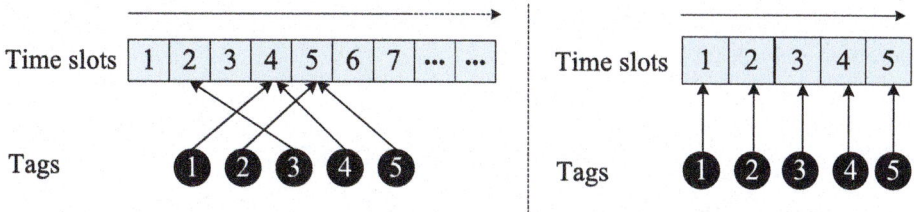

Fig. 6.3 Comparison of P2M (the left) and P2P (the right) for multiple tags. P2M would waste some slots that are collided (slots 4 and 5) or empty (slots 1, 3, 6, and 7). While P2P can selectively read tags and only needs five slots

6.5.1 Point-to-Point Selective Query

The Gen2 standard provides a command *Select* that allows the reader to selectively read a subset of tags based on user-defined criteria. As shown in Fig. 6.4, the selective query includes two phases: tags filtering and tag query. First, the reader issues a *Select* that specifies a bitmask and an action that will be performed by the tags. On receiving *Select*, each tag checks whether it matches the reader-to-tag bitmask. If yes, it will assert its flag variable SL; otherwise, it will deassert the SL. By carefully designing the bitmask, we can ensure only one tag can pass the bitmask comparison, which will be presented shortly. Then the reader further sends a *Query* that specifies the tags with asserted SL to reply. Since only one tag meets the requirement in P2P, this tag is the only one replying to the *Query* with its *RN16*. Subsequently, the reader transmits *ACK* with the decoded *RN16* and prepares to receive the *EPC* of this tag. When this query finishes, the reader will repeat the above process to read the tags one by one.

The desired property of P2P is its capacity to specify an individual tag to reply. If there is no response from this tag, the reader will know its absence. As a result, P2P can identify

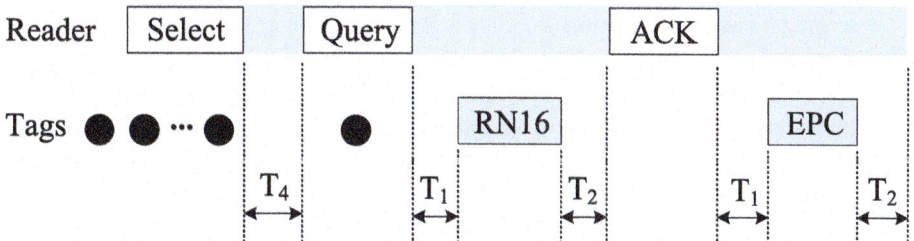

Fig. 6.4 Link timing of P2P communication where the black points represent tags. The Gen2 standard has strict requirements for each command format and link timing parameters T_1, T_2, and T_4 that stand for an interval-command time, enabling the computation of overall interrogation time

all m missing tags after n selective queries. Moreover, P2P can also detect a missing tag in at most $n - m$ selective queries.

6.5.2 Calculation of the Overall P2P Execution Time

As shown in Figs. 6.1 and 6.4, the length of a P2P selective query on a present tag contains a *Select*, T_4, a *Query*, and a singleton slot whose length is equal to that in P2M. If a missing tag is queried, the components of this query duration are almost the same as the prior except that slot duration becomes to empty slot size instead of singleton slot size. Thus, recall Sect. 6.4.4, we know that it takes P2P time of $(Select + Query + ACK) \cdot$ Tari $+ \frac{RN16+EPC}{BLF/j} + T_4 + 2(T_1 + T_2)$ to achieve a selective query on a present tag, where $j \in \{1, 2, 3, 4\}$ indicates different tag-to-reader encoding methods [4]. As a consequence, the overall time cost of P2P is $(n - m)\big((Select + Query + ACK)\text{Tari} + \frac{RN16+EPC}{BLF/j} + T_4 + 2(T_1 + T_2)\big) + m \cdot ((Select + Query) \cdot \text{Tari} + T_4 + T_1 + T_3)$.

Having described the process of P2P, we next explain how *Select*, the key function in P2P, is designed in our missing tag detection protocol.

6.5.3 *Select* Function

There are six mandatory fields in the *Select* command as shown in Fig. 6.5, we introduce five fields relevant to our design.

1. Action specifies eight types of tag behavior which are listed in Table 6.1. In our scenario, we use the first type, i.e., Action $= 000_2$, to specify tag action. Specifically, tags that match the received bitmask, called matching tags, will assert SL, while the other tags, called not-matching tags, will deassert SL.
2. MemBank indicates which tag memory model a tag will search to compare with the received bitmask. The MemBank $= 00_2$ is reserved memory storing passwords associated with the tag. If MemBank $= 01_2, 10_2, 11_2$ then the tag searches for the bitmask in the EPC memory bank that stores the tag EPC, TID memory bank that specifies the permalocked tag and manufacture specific information, and User memory bank that can

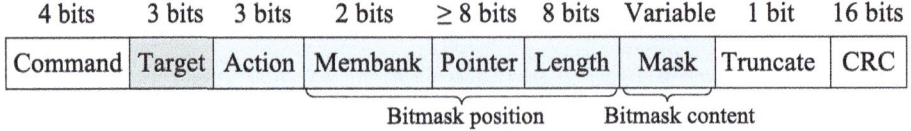

Fig. 6.5 *Select* command: MemBank, Pointer and Length specify the bitmask position that the tag needs to search in its memory; Mask records the bitmask content that the tag will compare with

6.5 P2P: Point-to-Point Missing Tag Identification

Table 6.1 Tag response to action

Action code	Tag matching	Tag not-matching
000_2	Assert SL	Deassert SL
001_2	Assert SL	Do nothing
010_2	Do nothing	Deassert SL
011_2	Negate SL	Do nothing
100_2	Deassert SL	Assert SL
101_2	Deassert SL	Do nothing
110_2	Do nothing	Assert SL
111_2	Do nothing	Negate SL

    be written with user-defined data. We employ the EPC memory bank in this chapter, i.e., MemBank $= 01_2$.

3. Pointer records a starting bit position in the chosen MemBank for the bitmask comparison.
4. Length specifies the bitmask length. If MemBank $= M$, Pointer $= p$ and Length $= l$ then the tag compares the bitmask with the bits starting from the p-th bit to the $(p+l-1)$-th bit in its memory model M.
5. Mask records the bitmask content that is a bit string. The tag compares it with the specified bit string in its memory.

From the description above, we observe that the combination of MemBank, Pointer and Length specifies the position of the bit string that the tag needs to search for in its memory while Mask records the bitmask content that the tag will compare with the bit string. Thus, we use BM to represent a bitmask, that is to say, $BM = (M, p, l, Mask)$.

In P2P, we build the bitmask from a tag EPC by setting MemBank $= 01_2$. The EPC is unique and has been stored in tags, thus P2P does not need to write new data to tags. We take an example to further illustrate its application. As shown in Fig. 6.6, the reader sends a *Select* specifying the EPC 1010 as the bitmask.[5] Upon receiving this command, each tag checks the bit string from the first to the fourth bit in its EPC and compares it with the received one in the Mask. Since only the gray tag meets the criterion, it will assert its SL and wait for the incoming *Query*, while the others will keep silent. We present an implementation of this example in Java in Fig. 6.7. As tag EPC starts from the 32nd bit in the memory, the pointer in the implementation is set to 0x20.

---

[5] Usually a Gen2 tag has a 96-bit EPC. In this example, we assume the EPC length is four for simplicity.

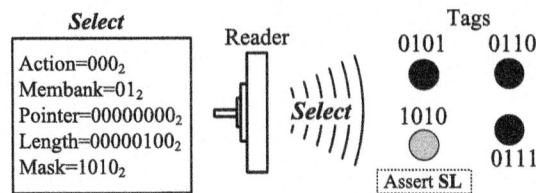

Fig. 6.6 Illustration of a selective query in P2P. There are four tags with EPCs: 0101, 0110, 1010, and 0111, respectively. With the configuration: `Action=000`$_2$, `Membank=01`$_2$, `Pointer=00000000`$_2$, `Length=00000100`$_2$, `Mask=1010`$_2$, the reader asks the tags to compare the bit string from the 1st to the 4th bit in their individual EPCs with the content in `Mask` of the received *Select*[5]

Fig. 6.7 Implementation of *Select* command in Fig. 6.6

```
TagFilter t1 = settings.getFilters().getTagFilter1();
    t1.setBitCount(4);
    t1.setBitPointer(0x20);
    t1.setMemoryBank(MemoryBank.Epc);
    t1.setFilterOp(TagFilterOp.Match);
    t1.setTagMask("A");
    settings.getFilters().setMode(TagFilterMode.OnlyFilter1);
```

So far we have introduced the framework of P2P and the *Select* function, the final question left is how to effectively and efficiently configure the bitmask, i.e., `Mask`. We attack the configuration of bitmask in the next subsection.

6.5.4 Bitmask Selection

Recall that in P2P, the reader seeks to distinguish a tag from the others in every slot. To do so, a direct way is setting `Mask` to the tag EPC, as the toy example in Fig. 6.6. Although such a configuration is effective, it suffers from low efficiency. Recall Fig. 6.5, a *Select* command is 45-bit long excluding the `Mask`.[6] If we use 96-bit EPC in `Mask` which is over two times of the other fields and over the two-thirds of the whole *Select*. If we can use a shorter `Mask`, the efficiency will be improved. For example, reconfiguring *Select* in Fig. 6.6 to `Pointer=00000000`$_2$, `Length=00000001`$_2$, `Mask=1`$_2$ when the tags compare the first bit of their EPCs with the `Mask`, we can make the gray tag the only one to meet the requirement with 1-bit mask instead of previous 4 bits.

Inspirited by the example above, we exploit the potential of building a bitmask with a portion of a tag EPC instead of the whole. Although 96~496-bit EPC can be supported by tags like ImpinJ Monza tags, we use 96-bit EPC in this chapter, but our work can be directly used in the scenarios where EPC length is over 96 bits. We know that 96-bit strings can

---

[6] The format of `Pointer` is an extensible bit vector that contains one or multiple 8-bit blocks. With one block, it can represent numeric values between 0 and 2^7. For the value over 2^7, it must add another block. Since the EPC length used in this chapter is 96 bits, it is enough to use one block, that is to say, field `Pointer` is 8-bit long.

6.5 P2P: Point-to-Point Missing Tag Identification

uniquely identify $2^{96} = 7.9 \times 10^{28}$ tags at most. Since the number of the tags in a system is usually much smaller than this quantity, the present EPCs in a Gen2 system are sparse compared with overall 2^{96} EPCs. We can exploit this sparsity to design more efficient bitmask selection methods. Note that their efficiency is more significant for tags with longer EPC, e.g., 496-bit EPC.

A deterministic Bitmask selection algorithm

We first design a deterministic algorithm, whose core idea is to use only a portion of a tag EPC as bitmask so that only one tag matches. The fields Length and Pointer specify the length and the starting position of the bit string in tag memory which will be compared with the received bitmask, we denote them by l and p, respectively. Since we select l consecutive bits from an $a \log n$-bit EPC, l could be equal to a value between 1 to $a \log n$, and there are $a \log n - l + 1$ segments in all in an EPC corresponding to $p = 0 : a \log n - l$. For instance, if $l = 2$ in Fig. 6.6, we have three segments for the gray tag from left to right, namely 10, 01, and 10. As a result, we can find an optimal bitmask in each slot, i.e., the shortest bitmask that can make a tag singular in a slot, through the following three-dimensional search (Algorithm 1). In the algorithm, $x(p, l)$ denotes a string from the p-th bit to $(p + l - 1)$-th bit in the EPC of tag x; $a = \frac{EPC}{\log n}$. The Algorithm, whose core steps are explained below, outputs the shortest bitmask specifying Pointer, Length and Mask.

- First, let $l = 1$, and we arbitrarily pick one out of n tags.
- Second, given l and this tag EPC, we compare its first l-bit segment, i.e., the leftmost, with those of the other $n - 1$ tags EPCs. If we find the segment unique, it can be used as a bitmask and Pointer$= 00000000_2$, then the searching process will be terminated; otherwise, this tag is regarded useless temporally, and we choose another one from the $n - 1$ tags to compare its first l-bit segment with those in the other $n - 1$ tags EPCs. This step runs until either a unique l-bit segment is found or any two tags has compared with each other.
- Third, if we fail to find a unique l-bit segment in the second step, we repeat the operations in the second step with the second l-bit segment. If it succeeds this time, this segment is assigned to Mask and Pointer is equal to 00000001_2; otherwise we set $l = l + 1$. The third step stops if a bitmask is found or $l = a \log n$. If a bitmask is found, that is to say, we can selectively query a tag matched this bitmask, then the algorithm keeps running to look for a bitmask for another tag.

From the description above, we can interpret the three dimensions in our algorithm as follows:

- Comparing between any two tags;
- Sliding Pointer p from 0 to $a \log n - l$;
- Incrementing l from 1 to $a \log n$.

Algorithm 1: Deterministic bitmask selection

Input : Tag set $\{x_1, x_2, \cdots, x_n\}$
1 **Initialization:** $l \leftarrow 1, N \leftarrow \emptyset, j \leftarrow 0, S^* \leftarrow \emptyset$
2 **while** $j \leq n$ **do**
3   **while** $l \leq a \log n$ **do**
4         $p \leftarrow 0$
5   **while** $p \leq a \log n - l$ **do**
6             $N \leftarrow \{x_1, x_2, \cdots, x_n\}; S \leftarrow \emptyset$
7   Indicator=1
8   Choose an arbitrary tag x from $N - S - S^*$
9   **for** *each* $j \in N/x$ **do**
10   **if** $x(p, l) == j(p, l)$ **then**
11                     $S \leftarrow S \cup x$; Indicator=0
12                     $p \leftarrow p + 1$; Jump to Line 6
13   **end**
14   **end**
15   **if** *Indicator==1* **then**
16   Record x, p, and l; $S^* \leftarrow S^* \cup x$
17                 $j \leftarrow j + 1$; Jump to Line 2
18   **else**
19                 $p \leftarrow p + 1$
20   **end**
21   **end**
22         $l \leftarrow l + 1$
23   **end**
24     $j \leftarrow j + 1$
25 **end**
26 Return a collection of $x(p, l)$

Our algorithm can deterministically find an optimal bitmask. We now analyze its computational complexity. As we explained previously, the complexity of our algorithm can be decomposed into three parts: 1) $\mathcal{O}(n^2)$ operations for each (p, l); 2) $\mathcal{O}(\log n^2)$ combinations of (p, l); 3) the algorithm needs to find a bitmask for all n tags. The overall computational complexity sums up to $\mathcal{O}(n^3 (\log n)^2)$.

A probabilistic Bitmask selection algorithm

We next devise a probabilistic Bitmask selection algorithm that ensures a unique bitmask with a required success probability. Compared with the deterministic algorithm, the probabilistic algorithm has three advantages:

- Reduced complexity. The probabilistic scheme reduces the complexity from $O(n^3)$ to $O(n^2)$ in the worst case. In practice the gain can be more significant. Low complexity is desired especially for handhold mobile readers which has limited computational capacity.

6.5 P2P: Point-to-Point Missing Tag Identification

- Tunable accuracy. As a desired property, the accuracy of the probabilistic algorithm can be tuned to strike a balance between the accuracy and computation and communication overhead.
- Better applicability. The probabilistic algorithm can be used to identify missing tags even when there are new tags that are not recorded in the database, but the deterministic one cannot conduct this task. This will be discussed at the end of this section.

In the probabilistic algorithm, we divide a tag EPC into $\lfloor \frac{|EPC|}{l} \rfloor$ non-overlapping segments, i.e., $\lfloor \frac{a \log n}{l} \rfloor$ segments. For example, if $l = 2$ in Fig. 6.6 where an EPC is assumed to be four bits long, we have two such segments for the black tag 0111 from left to right, namely 01 and 11. This method is formally stated in Algorithm 2 that operates as follows:

- First, we set l and another parameter z that stands for the execution rounds of this algorithm. How to configure the parameters will be introduced shortly.
- Second, we arbitrarily choose a tag and select the first (leftmost) segment of its EPC. Then, we compare this segment with those of the other $n - 1$ tags. If this segment is unique, we use it as a bitmask and set Pointer= 00000000_2, then the algorithm stops; otherwise, we select the second segment and repeat the operations above. The algorithm terminates when a unique segment is found or the number of the executed rounds exceeds z.

Algorithm 2: Probabilistic Bitmask selection

Input : Tag set $\{x_1, x_2, \cdots, x_n\}, l, z$

1 **Initialization:** $N \leftarrow \{x_1, x_2, \cdots, x_n\}, k \leftarrow 1, p \leftarrow 0$;
        choose an arbitrary tag x from N
2 **while** $k \leq z$ **do**
3   Indicator=1
4   **for** each $j \in N/x$ **do**
5   **if** $x(p, l) == j(p, l)$ **then**
6   Indicator=0
7   **end**
8   **end**
9   **if** *Indicator==1* **then**
10   Stop
11   **else**
12         $p \leftarrow p + l; k \leftarrow k + 1$
13   **end**
14 **end**
15 Return $x(p, l)$

It is obvious that the complexity of the probabilistic method is $\mathcal{O}(n \cdot z)$ where $z \leq \lfloor \frac{a \log n}{l} \rfloor$. To find bitmasks for all tags in P2P, this method needs to run n times, so the overall complexity is $\mathcal{O}(n^2 \log n)$.

Next, we move to the analysis of parameter configuration. Since each bit in EPC is generated randomly in practice, the strings of $\lfloor \frac{a \log n}{l} \rfloor$ non-overlapping segments are mutually independent. The algorithm would run k rounds if the first $k - 1$ rounds fail where $k \leq z$, thus the probability distribution of the number of executed rounds, defined as K, can be formulated as a geometric distribution.

Consider an arbitrary round, finding unique bitmasks for all n tags is equal to the event that the selected l-bit segments are different from each other. The probability of this event is $e^{-\frac{n^2}{2^{l+1}}}$ [23]. As a result, we have

$$\Pr(K = k) = (1 - e^{-\frac{n^2}{2^{l+1}}})^{k-1} \cdot e^{-\frac{n^2}{2^{l+1}}}.$$

Hence, the success probability after z rounds, defined as P_s can be calculated as

$$P_s = \sum_{k=1}^{z} (1 - e^{-\frac{n^2}{2^{l+1}}})^{k-1} \cdot e^{-\frac{n^2}{2^{l+1}}} = 1 - (1 - e^{-\frac{n^2}{2^{l+1}}})^z.$$

Denote by α the required success probability of finding bitmasks for n tags, we can get the relationship of l and z:

$$P_s = \alpha \implies \log(1 - \alpha) = z \log(1 - e^{-\frac{n^2}{2^{l+1}}}). \tag{6.1}$$

To solve this equation, we can first specify a value for either l or z, and derive the other. P_s monotonously increases with l and z, thus the selection of l and z indicates the trade-off between computational complexity and communication overhead. For example, let $z = 1$, the complexity will be reduced to $\mathcal{O}(1)$ while l reaches its maximum value $\log \frac{n^2}{-\ln \alpha} - 1$ from (6.1). If the required α is equal to 99% and $n = 2^{10}$, then $\log \frac{n^2}{-\ln \alpha} - 1 \approx 26$. In contrast, if let $z = \lfloor \frac{96}{l} \rfloor$ under the same requirement, we have $l = 20$ while $z = 4$. Note that the value of l cannot exceed the length of a tag EPC.

6.5.5 Missing Tag Identification with New Tags

In this part, we discuss whether P2P can be used to identify missing tags in the scenario with the arrival of new tags that are not recorded in the database. To do so, we study in two cases: P2P with the deterministic algorithm and P2P with the probabilistic algorithm.

In the first case, P2P cannot be used in such a coexistence scenario as the deterministic algorithm must search for all EPCs of the tags in the database to find a unique bitmask while those of the new tags are not recorded. As a result, some new tags may also match the selected bitmask, colliding with the response of the known tag, which makes P2P fail.

In the second case, P2P can be adapted for the coexistence scenario if the number of the new tags can be estimated or the reader knows the range of the new tag population. Assume the upper bound of the new tag populations is $\bar{u}$, given the required α, we can calculate the needed bitmask length l and the number of the execution rounds from the following equation such that the identification probability is at least α,

$$\log(1-\alpha) = z\log(1 - e^{-\frac{(n+\bar{u})^2}{2^{l+1}}}).$$

Note that when the tag EPC is 96-bit long, P2P can deterministically identify all missing tags if $l = 96$.

6.6 Implementation

6.6.1 Implementation Setup

COTS Gen2 devices: We use one ImpinJ R420 reader and 20 ImpinJ Monza-4 UHF tags in our implementation. These devices are completely compiled with the Gen2 standard. The missing identification programs are written in Java on the top of ImpinJ SDK v.1.28.0.1. In particular, the ImpinJ R420 reader supports Q-query and selective query. The ImpinJ Monza-4 tags have 96-bit EPCs.

Parameters: The transmission power of the reader is set to 30dbm, and its reception sensitivity is -70 dbm. We implement three tag-to-reader encoding methods: M2, M4, M8. As the ImpinJ reader can support three combinations, we vary the tag-to-read link rate from 320kbps with M2, to 68.5kbps with M4, to 21.33kbps with M8. In PMP, we set $Q = 2 \log n$ where n is the number of the tags in the Gen2 system, which will be set to 5, 10, and 20, respectively. We will investigate the correctness of the deterministic bitmask selection method and the probabilistic method, but use the former in the implementation of P2P while the latter will be used in the subsequent experiments where the system scales.

6.6.2 Implementation Results

We evaluate the performance of the proposed missing tag identification protocols, namely P2M and P2P. We would like to note that this chapter focuses on performance comparison in the same settings rather than maximizing the throughput.

Protocol investigation: Before the formal comparison, we first present how the deterministic bitmask selection method works. We start with $n = 5$ tags whose EPCs are listed in the

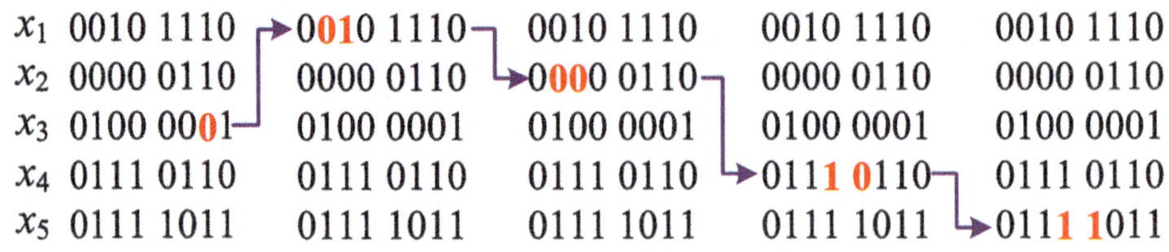

x_1 0010 1110   ⇢0010 1110⌐ 0010 1110   0010 1110   0010 1110
x_2 0000 0110   0000 0110 ⇢0000 0110⌐   0000 0110   0000 0110
x_3 0100 0001⇠   0100 0001   0100 0001   0100 0001   0100 0001
x_4 0111 0110   0111 0110   0111 0110   ⇢0111 0110⌐   0111 0110
x_5 0111 1011   0111 1011   0111 1011   0111 1011   ⇢0111 1011

Fig. 6.8 The bitmasks used in P2P. There are five tags and we present the first two words of EPCs in binary. we can first set the bitmask $BM = (000_2, 6, 1, 0_2)$ to query tag x_3, then use $BM = (000_2, 1, 2, 01_2)$, $BM = (000_2, 1, 2, 00_2)$, $BM = (000_2, 3, 2, 10_2)$, $BM = (000_2, 3, 2, 11_2)$ in sequence to query tags x_1, x_2, x_4, x_5, respectively

first column of Table 6.2, i.e., tags x_1 —x_5. Running Algorithm 1, we can first set the bitmask $BM = (000_2, 6, 1, 0_2)$ to query tag x_3, then use $BM = (000_2, 9, 1, 0_2)$, $BM = (000_2, 11, 1, 0_2)$, $BM = (000_2, 37, 1, 0_2)$, $BM = (000_2, 39, 1, 0_2)$ in sequence to query tags x_4, x_1, x_2, x_5, respectively. That is to say, it is sufficient for P2P to use a one-bit bitmask in this case. For illustration, we take a toy example where only the first two words of tag EPCs are searched, as shown in Fig. 6.8. Comparing this example with the prior, we can observe that searching more positions in EPC will yield shorter bitmasks.

We further execute Algorithm 1 to build the bitmasks for the cases of $n = 10$ and $n = 20$ corresponding to the first two columns and all tags in Table 6.2, respectively. The results for $n = 5, 10, 20$ are shown in Tables 6.3, 6.4, and 6.5, respectively. Note that we employ MemBank = 000_2 in P2P, and we just list $(p, l, Mask)$ for each tag for illustrative clarity.

Protocol comparison: From this part, we begin to compare the performance of P2M with P2P using the deterministic bitmask selection method in terms of execution time spent in identifying all missing tags and detecting the first missing tag under three different tag-to-reader encoding methods supported by an ImpinJ reader, namely M2, M4, and M8.

First, we investigate the impact of overall tag population n on the performance of P2M and P2P. To this end, we fix the number of missing tags $m = 0$ while increasing n from 5, to 10, to 20. As shown in Fig. 6.9, P2P can query all tags within significantly less time than P2M, and the performance gain soars with the increment in the number of tags in the system. Meanwhile, the execution time of P2M experiences more sharp increase than P2P does. For example in Fig. 6.9a, when the tag population is 5, P2P is 1.5× better than P2M.

Table 6.2 Tag EPCs in the implementation

i	x_i	x_{i+5}	x_{i+10}	x_{i+15}
1	2E4E6693572D3A8D185E0988	110B1D467E616FCA07E03A31	6402201E11FA2CB336243D3A	29B66F4D3EBD748A42352298
2	06DD7F27437B193326BA3F35	70A575FE134C343C67F778CA	37A721130D0879BC3BAA253E	3636306E7A131BFF738758C6
3	415859552FF64559679B4EFE	300833B2DDD9140000000000	4EB922210CEF339B2B3C0F4B	2FE666A910E74FB543FE5D83
4	76317A5F05056B4072D21075	49D87D2252B13F24278A24CF	75643B7A0D806EA8286E08BD	22A03BE81F5F28F552EF2011
5	7BD8536F240C0F0C19C2534A	2E8B6D541CCD447E0B7C684D	57EA364D50A277C53EB21B13	1B48018C6AB05C2274F13B9F

6.6 Implementation

Table 6.3 Bitmasks for $x_1 - x_5$

i	x_i
1	$(11, 1, 0_2)$
2	$(37, 1, 0_2)$
3	$(6, 1, 0_2)$
4	$(9, 1, 0_2)$
5	$(39, 1, 0_2)$

Table 6.4 Bitmasks for $x_1 - x_{10}$

i	x_i	x_{i+5}
1	$(13, 2, 11_2)$	$(2, 2, 01_2)$
2	$(45, 2, 01_2)$	$(8, 2, 10_2)$
3	$(21, 2, 00_2)$	$(32, 1, 1_2)$
4	$(10, 2, 11_2)$	$(40, 2, 10_2)$
5	$(3, 2, 11_2)$	$(19, 2, 01_2)$

Table 6.5 Bitmasks for $x_1 - x_{20}$

i	x_i	x_{i+5}	x_{i+10}	x_{i+15}
1	$(11, 3, 011_2)$	$(7, 3, 100_2)$	$(1, 3, 110_2)$	$(33, 3, 110_2)$
2	$(1, 3, 000_2)$	$(6, 3, 001_2)$	$(13, 3, 111_2)$	$(35, 3, 01_2)$
3	$(20, 3, 100_2)$	$(4, 4, 1111_2)$	$(52, 2, 00_2)$	$(2, 2, 01_2)$
4	$(11, 3, 100_2)$	$(74, 3, 001_2)$	$(5, 3, 101_2)$	$(4, 3, 001_2)$
5	$(70, 2, 01_2)$	$(55, 3, 001_2)$	$(68, 2, 11_2)$	$(18, 2, 00_2)$

While this number increases to 5× when there are 20 tags. The performance gain of P2P comes from the point-to-point design as it is able to successfully read a tag in every slot, but it takes $\mathcal{O}(n)$ slots for P2M to access a tag on average.

Second, we move to study how P2M and P2P perform under different missing tag populations in the system. To do so, we fix the number of overall tags $n = 20$ while changing the number of missing tags m as $m = 4, 6, 8, 12$. The experimental results are depicted in Fig. 6.10. From these results, we can observe the following phenomenons:

- Overall performance: P2P remarkably outperforms P2M. Specifically, the identification cost of PMP, as shown in Fig. 6.10a, falls into the range between 0.56 s and 1.49 s, which is 2.8× to 5.4× more than that of P2P. In the other tag-to-reader rates, P2P achieves at least 3.3× performance gain over P2M. This is primarily due to the point-to-point query paradigm that reads tags in sequence while P2M needs more time to tackle collisions.
- Impact of missing tags: As the number of missing tags increases, the execution time of P2M decreases more significantly than P2P. For instance in Fig. 6.10a, the reduction of

Fig. 6.9 Performance comparison with different numbers of overall tags under three tag-to-reader rates: M2 (320kbps) > M4 (68.5kbps) > M8 (21.33kbps)

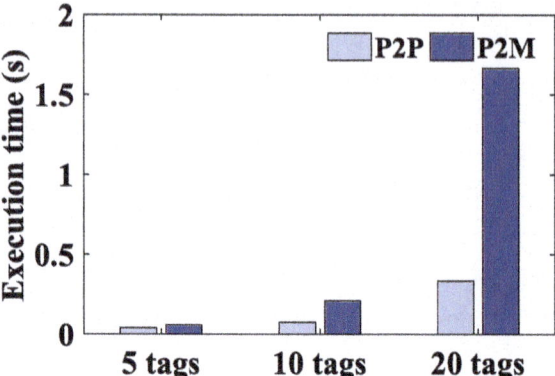

(a) Tag-to-reader encoding method: M2

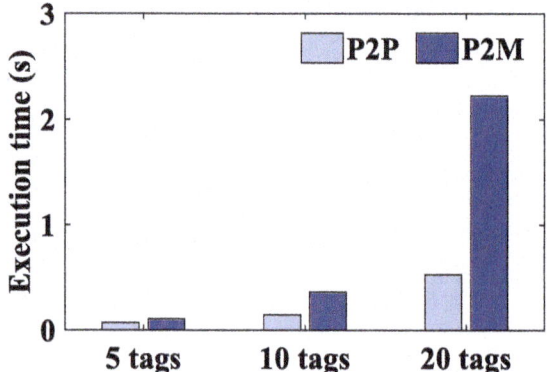

(b) Tag-to-reader encoding method: M4

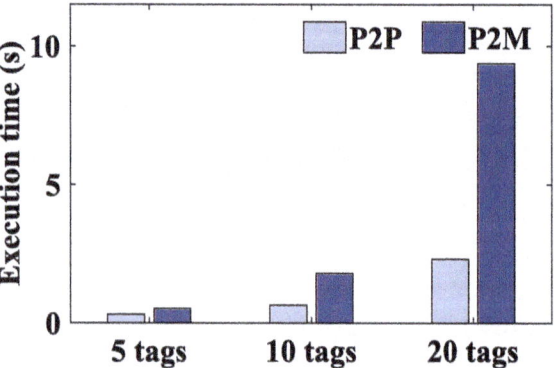

(c) Tag-to-reader encoding method: M8

6.6 Implementation

Fig. 6.10 Performance comparison with different missing tag population under three tag-to-reader rates: M2 (320kbps) > M4 (68.5kbps) > M8 (21.33kbps)

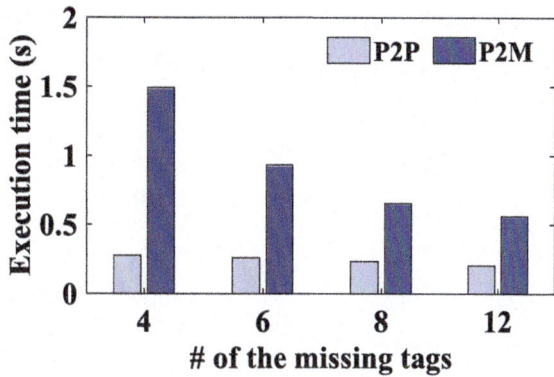

(a) Tag-to-reader encoding method: M2

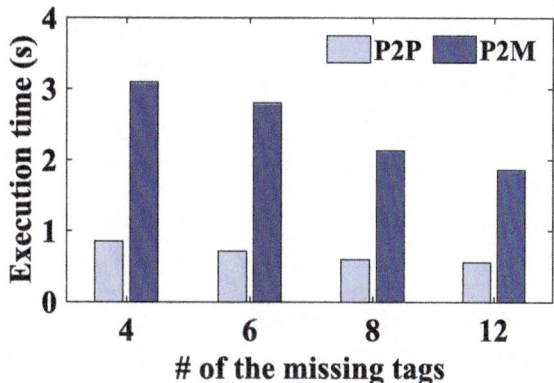

(b) Tag-to-reader encoding method: M4

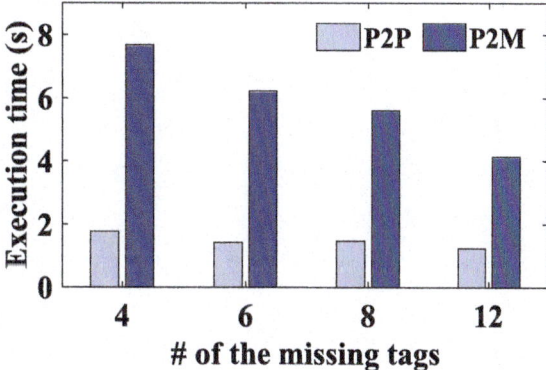

(c) Tag-to-reader encoding method: M8

Fig. 6.11 Performance comparison in terms of detection time indicating the time of finding the first missing tag

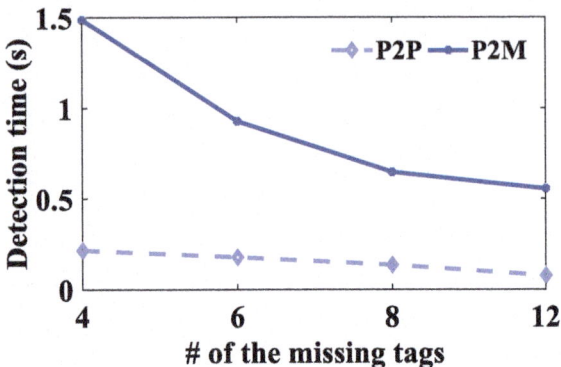

(a) Tag-to-reader encoding method: M2

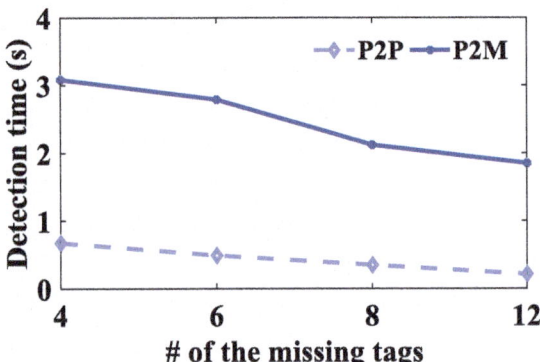

(b) Tag-to-reader encoding method: M4

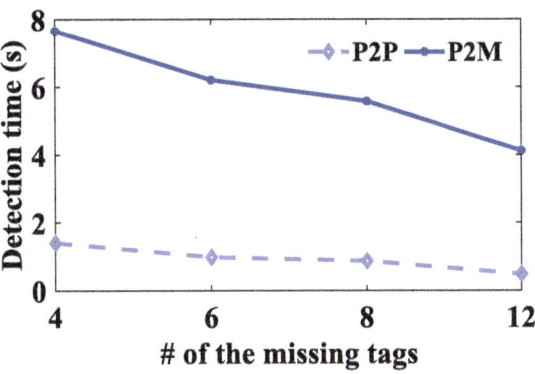

(c) Tag-to-reader encoding method: M8

6.6 Implementation

Table 6.6 Bitmask length l and execution rounds z when $n = 5$

α \ l	3	4	5	6	7	8–10	11	12	13	14
0.99	20	8	5	3	2	2	1	–	–	–
0.999	30	12	7	4	3	3	3	3	3	1

P2M is 62.2%, which is 2.4 times that of P2P. This can be interpreted as follows: the increase of missing tags reduces tag collisions in P2M but has a lower impact on P2P as it employs point-to-point queries.

Under the same settings as the above, we further compare P2M and P2P in terms of missing tag detection time that is the time spent in finding the first missing tag. It can be observed from Fig. 6.11 that P2P is able to detect the first missing tag within quite less time than P2M. In particular, When there are 12 missing tags, it takes P2M with M2 nearly 7× time as much as P2P to find the first missing tag. The performance gap between them reaches over 8× when M4 and M8 are used. Look at Figs. 6.10 and 6.11, we can also find that the detection time of P2P significantly reduces especially in the presence of more missing tags while that of P2M does not change. This difference is resulted from the nature of P2P and P2M in that the former can learn the existence or absence of a tag in each slot but the latter cannot know tag states until the execution of the whole frame. That said, P2P can find a missing tag after probing $n - m$ tags in the worst case while P2M is expected to query all n tags.

Correctness of the probabilistic bitmask selection method: Having implemented P2M and P2P with 20 ImpinJ tags, we move to confirm the correctness of the probabilistic bitmask selection method in this part. Revisiting Table 6.2 where the 20 tag EPCs are listed, we first check whether Algorithm 2 works in the 5-tag scenario. To assess the reliability of the probabilistic method, we set $\alpha = 0.99$ and 0.999, and run Algorithm 2 for $\frac{1}{1-\alpha}$ times. Each time we randomly select 5 out of 20 tag EPCs. If bitmask collisions among tags arise more than one time, we claim the failure of Algorithm 2. We record in Table 6.6 the combinations of l and z that fulfill the required probability. The results show that with α increased Algorithm 2 needs to use a longer bitmask or run more rounds, which is in correspondence with the analytical results. Moreover, given an α, the increase of either l or z can yield a smaller value of the other, confirming the tradeoff between communication overhead and computational complexity.

To evaluate the impact of the system scale, we increase the number of tags from 50 to 300 with a step length of 50 and generate tag EPCs at random. From the results, we observe that Algorithm 2 can achieve the required probability α with the tag population increased. Since the maximum bitmask length can be directly computed from (6.1) with $z = 1$, we

Table 6.7 Bitmask length l and execution rounds z: (l, z)

α \ n	50	100	150	200	250	300
0.99	(11, 6)	(13, 6)	(15, 4)	(15, 6)	(16, 5)	(17, 4)
0.999	(12, 6)	(14, 7)	(15, 6)	(16, 6)	(16, 5)	(18, 5)

Table 6.8 Execution time of P2M and P2P with FM0

Protocol	γ	500	1,000	2,000	4,000	10,000
P2M	0.3	29.86	119.10	475.70	1901.40	1,1878.48
	0.6	29.78	118.94	475.37	1,900.76	1,1876.87
P2P	0.3	0.79	1.60	3.26	6.62	16.92
	0.6	0.72	1.46	3.00	6.06	15.52

Table 6.9 Execution time of P2M and P2P with M4

Protocol	γ	500	1,000	2,000	4,000	10,000
P2M	0.3	44.17	175.84	701.67	2,803.31	1,7508.01
	0.6	44.00	175.48	701.00	2,801.93	1,7505.00
P2P	0.3	1.085	2.19	4.44	9.0	22.82
	0.6	0.91	1.84	3.72	7.55	19.24

only list the combinations of the minimum bitmask length and execution rounds that make Algorithm 2 successful in Table 6.7. We can find that either bitmask length or execution rounds increase when the system scales or the required success probability becomes higher, which corresponds to the analytical result.

Performance evaluation under larger systems. We further show how the time efficiency of the proposed protocols changes as the system scales up. To this end, we set parameters following the Gen2 standard and specification of ImpinJ reader as follows: Tari= $12.5\mu s$, BLF= 640kHz. We use FM0 and M4 as encoding methods for the tag-to-reader link, respectively. Accordingly, the data rate defined by r is $1/BLF$ and $4/BLF$. The time durations are $T_1 = T_3 = \max(2.75Tari, 10r)$, $T_2 = 3r$, and $T_4 = 5.4r$. We vary the number of the overall tags from 500 to 10,000 and set $\alpha = 0.999$ when the required bitmask length l is 27, 29, 31, 33, 36 and the execution round of the probabilistic algorithm z equals to 1. In addition, define γ as the ratio of the number of the missing tags to that of the overall tags, we set it to 0.3 and 0.6. We listed the results in Tables 6.8 and 6.9.

We can observe that the increment in the execution time of P2M follows a square pattern of that in the number of the overall tags. The pattern becomes linear in P2P. Consequently,

P2P is considerably more time-efficient than P2M. We can also find that the ratio γ of the missing tag population has more impact on P2P than P2M. This is because the increase of γ leads to fewer success slots and more empty slots in P2P. An empty slot is shorter than a success slot. Yet due to the change of empty slots resulted from the increase of γ in P2M, which is in the order of magnitude $\mathcal{O}(n)$, is significantly smaller than the original number of empty slots, i.e., $\mathcal{O}(n^2)$.

6.7 Conclusion

In this chapter, we have proposed two protocols enabling the missing tag identification service with COTS RFID reader and tags. Specifically, we first used Q-query to develop a point-to-multipoint protocol that operates in an analog frame-slotted Aloha paradigm to collect tag EPCs. A missing tag can be found if the collected EPC set does not contain its EPC. We then devised a point-to-point protocol that employs a bitmask to specify one tag to reply in each slot so that tag response collisions are avoided and time efficiency is improved. Moreover, we presented two bitmask selection methods to build compact bitmasks. The proposed protocols were implemented in ImpinJ readers and tags, and the extensive results showed that they were able to achieve the missing tag identification task.

References

1. EPCglobal Inc., Class-1 generation-2 UHF RFID protocol for communications at 860 mhz–960 mhz (2005). http://www.gs1.org
2. ImpingJ Inc., Impingj connectivity devices. https://www.impinj.com/
3. Thingmagic Inc., Thingmagic products. http://www.thingmagic.com/index.php
4. Chain Store Age, Retailers losing billions to inventory shrink (2017). https://nrf.com
5. C. C. Tan, B. Sheng, Q. Li, How to monitor for missing RFID tags, in *IEEE ICDCS* (2008), pp. 295–302
6. W. Luo, S. Chen, T. Li, Y. Qiao, Probabilistic missing-tag detection and energy-time tradeoff in large-scale RFID systems, in *ACM MobiHoc* (2012), pp. 95–104
7. W. Luo, S. Chen, Y. Qiao, T. Li, Missing-tag detection and energy-time tradeoff in large-scale RFID systems with unreliable channels. IEEE/ACM TON **22**(4), 1079–1091 (2014)
8. M. Shahzad, A. X. Liu, Expecting the unexpected: Fast and reliable detection of missing RFID tags in the wild, in *IEEE INFOCOM* (2015), pp. 1939–1947
9. J. Yu et al., Finding needles in a haystack: Missing tag detection in large rfid systems. IEEE TCOM **65**(5), 2036–2047 (2017)
10. J. Yu, L. Chen, R. Zhang, K. Wang, On missing tag detection in multiple-group multiple-region rfid systems. IEEE TMC **16**(5), 1371–1381 (2017)
11. L. Yang, Q. Lin, C. Duan, Z. An, Analog on-tag hashing: Towards selective reading as hash primitives in gen2 rfid systems, in *ACM MobiCom* (2017), pp. 301–314
12. T. Li, S. Chen, Y. Ling, Identifying the missing tags in a large RFID system, in *ACM MobiHoc* (2010), pp. 1–10

13. R. Zhang, Y. Liu, Y. Zhang, J. Sun, Fast identification of the missing tags in a large RFID system, in *IEEE SECON* (2011), pp. 278–286
14. X. Liu et al., A multiple hashing approach to complete identification of missing rfid tags. IEEE TCOM **62**(3), 1046–1057 (2014)
15. X. Liu, K. Li, G. Min, Y. Shen, A.X. Liu, W. Qu, Completely pinpointing the missing RFID tags in a time-efficient way. IEEE TC **64**(1), 87–96 (2015)
16. Y. Zheng, M. Li, P-mti: Physical-layer missing tag identification via compressive sensing. IEEE/ACM TON **23**(4), 1356–1366 (2015)
17. M. Chen, J. Liu, S. Chen, Y. Qiao, Y. Zheng, Dbf: A general framework for anomaly detection in rfid systems, in *IEEE INFOCOM* (IEEE, 2017), pp. 1–9
18. K. Finkenzeller, *RFID Handbook* (John Wiley & Sons, 2010)
19. M. Feldhofer, C. Rechberger, A case against currently used hash functions in rfid protocols, in *International Conferences On the Move to Meaningful Internet Systems* (Springer, 2006), pp. 372–381
20. C. Rolfes, A. Poschmann, G. Leander, C. Paar, Ultra-lightweight implementations for smart devices–security for 1000 gate equivalents, in *International Conference on Smart Card Research and Advanced Applications* (Springer, 2008), pp. 89–103
21. Systemid, Immediate inventory management: Everyone wins with rfid technology at walmart (2012). http://www.systemid.com/learn/
22. RFID Journal, Rfid technology is boosting sales and customer engagement for retailers (2017). https://www.raconteur.net/business/
23. M. Mitzenmacher, E. Upfal, *Probability and computing: Randomized algorithms and probabilistic analysis* (Cambridge University Press, 2005)
24. K. Beyer et al., On synopses for distinct-value estimation under multiset operations, in *ACM SIGMOD* (2007), pp. 199–210

7 Conclusion and Perspective

In this chapter, we begin by summarizing the book's examination of the efficient utilization of RFID systems for missing event detection. Following this, we engage in a discussion on various unresolved questions and outline several potential avenues for future research, including energy utilization, anonymity, and tag implementation.

7.1 Book Summary

RFID technology, which employs low-power radio waves for the automatic identification of tagged objects and the retrieval of associated data, represents a superior alternative to optical barcodes that require significant manual intervention. As a result, RFID systems boast a broad spectrum of applications, including ticketing services, Electronic Toll Collection (ETC), library management, object tracking, logistics, industrial automation, and medical devices. Therefore, RFID technology emerges as a pragmatic solution for the IoT.

In this book, we delve into the fundamental applications of efficient and secure backscatter networks with RFID technology, with particular emphasis on the specialized domain of missing event detection. Readers can track items by querying the tags attached to physical objects within the workplace. The responses from these tags indicate the presence of objects. Moreover, using universal tags presents a lower cost solution that meets the growing demand for multi-tasking capabilities. Consequently, the challenge of missing tags detection revolves around efficiently grouping the tags and then swiftly scheduling non-responsive tags. In essence, the primary objective of developing efficient missing detection algorithms is to optimize tag accessibility for readers. However, designing such efficient missing detection algorithms faces several challenges arising from privacy concerns, complexities associated with multi-tagged scenarios, limited computational capacity on the tag side, unreliable wireless communication channels, and the pervasive yet intrusive nature of these tags.

These challenges highlight the critical need for innovative approaches to missing event detection algorithms, especially in scenarios involving multiple tagged objects. In this regard, we propose a systematic analysis and design based on our insights, aiming to provide new perspectives for future research on missing detection within RFID systems.

In particular, we commence by elucidating group labeling in Chap. 2, which serves as a fundamental management technique for RFID systems. We present an approximation algorithm accompanied by a proven competitive ratio, followed by the development of two streamlined algorithms characterized by reduced complexity while maintaining comparable performance. These advancements effectively address the NP-hard seed assignment problem associated with the utilization of multiple seeds, thereby enhancing efficiency in labeling tags and enabling group-wise management. Subsequently, Chap. 3 explores anonymous group writing to bolster the security of backscatter RFID systems. We construct an approximately random sequence as noise by overlapping transmission data from different tag groups, thus hiding the original information with low computational complexity. Chapters 4 and 5 introduce a series of algorithms aimed at detecting missing events within multi-tagged contexts, focusing on filter constructing and hash seed searching, respectively. In Chap. 4, we employ a filter to designate one tag attached to each object, facilitating access to the reader. We enhance the efficiency of this process through compression of the constructed filter, further optimizing broadcasting time efficiency. In Chap. 5, we propose a framework for missing detection that highlights trade-offs between computation and communication while considering characteristics of hash functions alongside time discrepancies between seed searching and communication. Finally, Chap. 6 tackles the challenge of missing tag identification using COTS tags, bridging theory with practical implementation. Initially, We used Q-query to develop a point-to-multipoint protocol and devise a point-to-point protocol employing a bitmask to specify one tag for response in each slot. The proposed protocols were implemented in COTS Gen2 reader and tags, thereby validating their effectiveness in achieving the missing tags detection.

In the subsequent section, we examine a range of open questions that we consider pertinent to the topics addressed in this book and outline several significant potential avenues for future research.

7.2 Open Questions and Future Work

7.2.1 Energy Utilization

The active tags, which are powered by batteries, provide enhanced functionality and extended communication ranges. However, they come with high production and management costs. In contrast, passive tags operate without batteries and enable large-scale deployment due to

7.2 Open Questions and Future Work

their cost-effectiveness. Nevertheless, the stringent energy limitations significantly restrict the functionality of these tags and limit their communication ranges. As a result, energy efficiency becomes a critical concern for RFID systems. In this context, the optimal solution lies in increasing energy income while simultaneously minimizing energy expenditures for passive tags.

A critical research avenue for enhancing energy income involves the effective harvesting of environmental energy. Intuitively, passive tags can extract energy from solar, wind, kinetic, and geothermal sources. However, these energy sources are profoundly influenced by surrounding environmental conditions, necessitating the incorporation of bulky energy harvesting devices (larger devices yield greater amounts of energy), which conflicts with the goal of miniaturization. Radiofrequency (RF) energy harvesting offers a promising solution, enabling tags to capture ambient RF energy from their environment. This method inherently provides resilience against climatic variations and facilitates miniaturization (the size of the RF antenna can decrease as the RF frequency increases). In urban environments, ambient RF energy is abundant due to the widespread deployment of digital TV (DTV) towers, cellular communication networks, and Wi-Fi hotspots. Therefore, it is essential to consider hybrid approaches that integrate energy collection from these diverse sources, taking into account the differing energy spectral densities between urban and suburban settings as well as the varying frequency bands of the aforementioned sources. Moreover, the rapid deployment of 5G networks has catalyzed innovations in the design of millimeter-wave-enabled power solutions, thereby opening new avenues for the utilization of 5G millimeter-wave energy in RFID systems.

On the other hand, the primary energy-intensive operations associated with tags involve data processing and RF signal transmission. The clock frequency of the processor dictates the processing rate. Generally, a higher frequency is correlated with quicker response times and increased data throughput. However, this also leads to greater power consumption. In RFID systems, elevated frequencies may not yield faster responses due to the intermittent nature of processing execution. This phenomenon arises because the energy required at elevated frequencies can surpass the stored energy available, necessitating that tags engage in cyclic energy harvesting during processing. A viable solution involves identifying an optimal frequency that minimizes processing time. Exploring this optimal frequency in real time, both prior to and during task execution, ensures that each process operates at its most efficient frequency. Nevertheless, this approach requires additional time for exploring frequency scaling and consumes extra computational resources, both of which contribute to overall energy expenditure. Therefore, a promising direction lies in quantifying the relationship between processing requirements and frequency selection, such quantification would facilitate pre-calculations aimed at mitigating energy overhead. Consequently, it is imperative to consider methods for rapidly selecting the optimal frequency with minimal power consumption. From the perspective of RF signal transmission, a straightforward strategy is to reduce both the number of responses generated and their duration. Thus, a critical direction is to explore novel methods for effectively managing the trade-off between communication time and energy costs.

7.2.2 Anonymity

With the widespread implementation of RFID systems, the volume of transmitted data has experienced significant growth, highlighting an urgent need for data anonymity to bolster security and privacy. Traditional methods depend on authentication protocols to prevent unauthorized users from eavesdropping and compromising data integrity. However, conventional mobile authentication protocols are ill-suited for RFID systems due to their limitations in energy and computational resources. As a result, a pivotal focus must be directed toward the development of efficient, ultra-lightweight authentication protocols. Moreover, contemporary authentication protocols require tags to be equipped with specialized encryption and decryption modules to ensure data anonymity, which consequently increases computational complexity. Additionally, these protocols typically necessitate the interrogation of each tag individually, which proves to be time-inefficient, particularly within large-scale systems. In this book, we present an ultra-lightweight encryption method utilizing logical operators aimed at enhancing the anonymity of group data through effective encryption techniques. Therefore, a promising approach is to integrate authentication-free mechanisms with ultra-lightweight encryption utilizing logical operators, thereby enhancing time efficiency while safeguarding data anonymity.

7.2.3 Tag Implementation

The passive tags have a vast application market owing to their low maintenance costs and prolonged system longevity, making them suitable for deployment in harsh environments and remote locations where active tags may be impractical. Traditional tags operate on the same frequency for both uplink and downlink transmissions, which can result in challenges such as self-jamming, multipath scattering, and suboptimal performance in cluttered conditions. In contrast, harmonic RFID tags function at two distinct frequencies—receiving signals at the fundamental frequency while backscattering at the harmonic frequency. This design imparts a significant degree of robustness against clutter and multipath interference. Consequently, an important area of focus is the development of a harmonic generator that operates with low energy consumption while ensuring miniaturization. Additionally, harmonic RFID systems necessitate two frequency duplexes for communication, requiring a substantial allocation of frequency bandwidth for large-scale deployment. Unfortunately, frequency spectrum resources are finite and often subject to licensing requirements. One potential solution to address this challenge is the adoption of multi-band communication. Specifically, the reader transmits using multiple ISM frequencies to the tags, which then

7.2 Open Questions and Future Work

mix and filter out the desired frequency band for responses. This approach shifts attention toward exploring interference cancellation techniques.

Further cost reduction of passive tags can be achieved through the elimination of chips, leading to the development of what are known as chipless tags. However, the absence of chips complicates data processing and communication. Specifically, anti-collision algorithms designed for chip-based systems cannot be directly applied to chipless architectures. A viable solution involves distinguishing each tag by its signal characteristics, employing techniques such as correlational signal processing in the time domain and frequency division access in the frequency domain. Nevertheless, challenges persist when dealing with a large number of tags. The read distance of a chipless tag is significantly shorter compared to that of a chip-based tag, primarily due to the lack of energy management inherent in chip technology. One potential remedy is to minimize background noise and interference. To achieve this goal, utilizing antenna arrays in the reader can enhance gain. Additionally, self-interference cancellation techniques may be employed to amplify the backscatter signal. Moreover, the data storage capacity within chipless tags is limited. There, it is crucial to focus on enhancing bit capacity within these tags.

The manufacturer's authorised representative in the EU is Springer Nature Customer Service Centre GmbH, Europaplatz 3, 69115 Heidelberg, Germany. If you have any concerns regarding our products, please contact ProductSafety@springernature.com

Printed and bound by CPI Group (UK) Ltd, Croydon, CR0 4YY
26/03/2026
02078941-0017